S0-AIW-878

# MAPPING THE WORLD
## The Story of Cartography

BEAU RIFFENBURGH

THE ROYAL
GEOGRAPHICAL
SOCIETY

CARLTON
BOOKS

*In memory of David Boss –
mentor, role model and
dear friend.*

*Author's Acknowledgements*

This book benefited from the knowledge of many experts
on cartography, for which I am most appreciative. I also
wish to express my thanks to Vanessa Daubney and Katie
Baxendale of Carlton Publishing Group, Bill Hamilton of
A. M. Heath & Co, and designers Liz Wiffen and Drew
McGovern for their invaluable contributions to this book.
I also am grateful for the editorial input given throughout
this project by Dr Liz Cruwys.

This is an Andre Deutsch book

Text © Wildebeest Publishing Ltd 2011, 2014
Design © Carlton Books Limited 2011, 2014

The right of Beau Riffenburgh to be identified as the
Proprietor of this Work has been asserted in accordance
with the Copyright, Designs and Patents Act 1988

This edition published in 2014 by Andre Deutsch Limited
A division of the Carlton Publishing Group
20 Mortimer Street
London
W1T 3JW

This book is sold subject to the condition that it shall
not, by way of trade or otherwise, be lent, resold, hired
out or otherwise circulated without the publisher's prior
written consent in any form of cover or binding other
than that in which it is published and without a similar
condition including this condition, being imposed upon the
subsequent purchaser.

Printed in China

All rights reserved
Previously published as *The Men who Mapped the World*

A CIP catalogue for this book is available from the
British Library

ISBN: 978 0 233 00439 6

# Contents

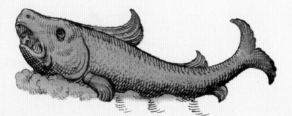

# Introduction

IN 1821 MANY OF THE GREATEST SCIENTIFIC MINDS IN EUROPE – including Pierre-Simon Laplace, Jean-François Champollion, and Alexander von Humboldt – helped found the Société de Géographie in Paris, the world's first modern geographical society. In short order, it was followed by the founding of the Gesellschaft für Erdkunde zu Berlin (1828), the Royal Geographical Society (RGS, 1830), and the Imperial Russian Geographical Society (1845).

A primary objective of these organizations was, as the Russian charter stated, to "collect, process and disseminate geographical, ethnographic and statistical information". The most conspicuous way this was done in the following decades – particularly by the RGS – was the promotion and sponsorship of geographical exploration, such as that by David Livingstone, Richard Francis Burton and Ernest Shackleton. Another way to attain this goal was to encourage the production of maps, particularly of newly discovered areas.

Maps were not a new phenomenon – they had been produced in numerous formats and for many different reasons for thousands of years. However, the geographical societies helped establish the first broad-based public interest in them, which fulfilled the same goal as maps do today: creating visual representations to help people define, describe or navigate through geographical or political regions, spaces real or imagined, or thematic issues. The maps of the time also followed the scientific requirements of modern maps, unlike some from earlier periods. They tended to do away with excess ornamentation, such as the personification of the winds or the inclusion of mythological creatures. As they do today, they contained a "scale" – a ratio comparing the size of a measured feature to its smaller representation – and an "orientation" to show its relationship with compass directions. Maps of large regions also included a "projection" – a system of mathematically definable rules designed to assist the transfer of three-dimensional information, such as that about the spherical Earth, onto a flat surface in a way to avoid distortions of shape, area, distance, direction or scale.

Soon, cartography – which can be defined as the scientific, technical and aesthetic practice of making maps, charts and plans – seemed for some to be the primary focus of geographical discovery. "The end result and final goal of all geographic investigations, explorations, and surveys is the depiction of the Earth's surface: the map," wrote the famous German geographer August Petermann.

Above: A map of northern Europe and Scandinavia from Battista Agnese's Portolan Atlas of 1553. Unlike many of the maps in the Atlas, which include rhumb lines and a compass rose, this map instead features an assortment of mythological sea monsters.

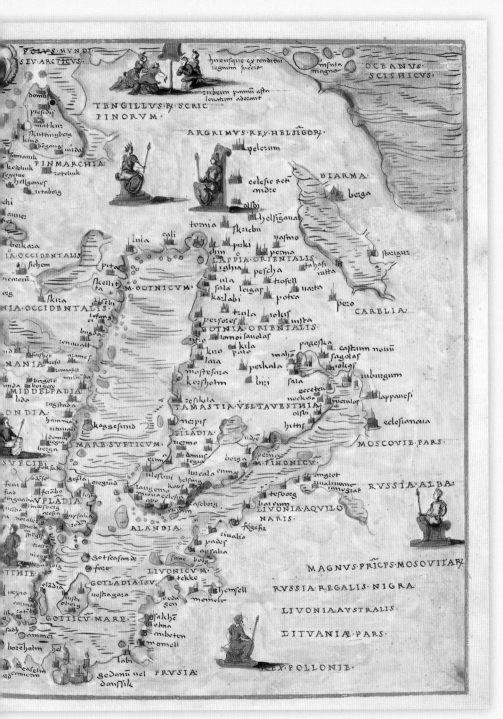

"The map is the basis for geography. The map shows us what we know about our Earth in the best, clearest and most exact way."

Petermann's emphasis on the relationship between geography and maps was not the only driving force historically for their production. Others included contributing to the rule and administration of society; determining political boundaries; aiding navigation; expressing personal or national wealth; conveying, suppressing, or falsifying new information; and producing propaganda. Or maps could just be the key to a ripping yarn, as they were for *King Solomon's Mines*, *Treasure Island* and *The Count of Monte Cristo*.

Today, maps are still produced for many reasons, but with an even wider variety of content. Among the incentives for making maps are locating sites on the surface of the Earth, highlighting relationships between factors, and showing patterns of distribution, whether those be political, geographical, geological, biological, historic, medical, hydrographical, administrative, or travel-related. The last of these is undoubtedly the most common, and vast numbers of people use maps on an almost daily basis, whether they are for following roads; finding correct seating at a sports stadium, cinema or airplane; or locating a particular store in a mall.

Such maps do not perhaps incorporate the aesthetic values and artistic talents of the medieval European *mappae mundi* with their flowing script and ornamentation or of the highly stylised Islamic works of the Balkhi school. But all of the maps presented in this book are worthy of note and study, from the earliest to the most recent. Sadly, countless pre-modern maps do not exist in their original form, although many were copied in versions that were often equally remarkable. Most importantly, because of the differing historical contexts, cartographic principles and perceived goals of the maps illustrated here, each tells a unique story and gives its own stimulating, alluring, and frequently beautiful, perception of a once-upon-a-time "reality".

**Beau Riffenburgh**

# CHAPTER 1

# Mapping in the *Ancient* and *Medieval World*

Right: Ptolemy's famous world map (c. AD150) was reproduced many times throughout medieval Europe. This version was prepared by Nicolaus Germanus around 1468–71. After being printed in Ulm, Germany, in 1482, the same wood blocks were used for an atlas published in 1486 by Johannes Reger. This copy of the atlas was later owned by the English textile designer and artist William Morris. Note that most of the Indian subcontinent is shown as a large island.

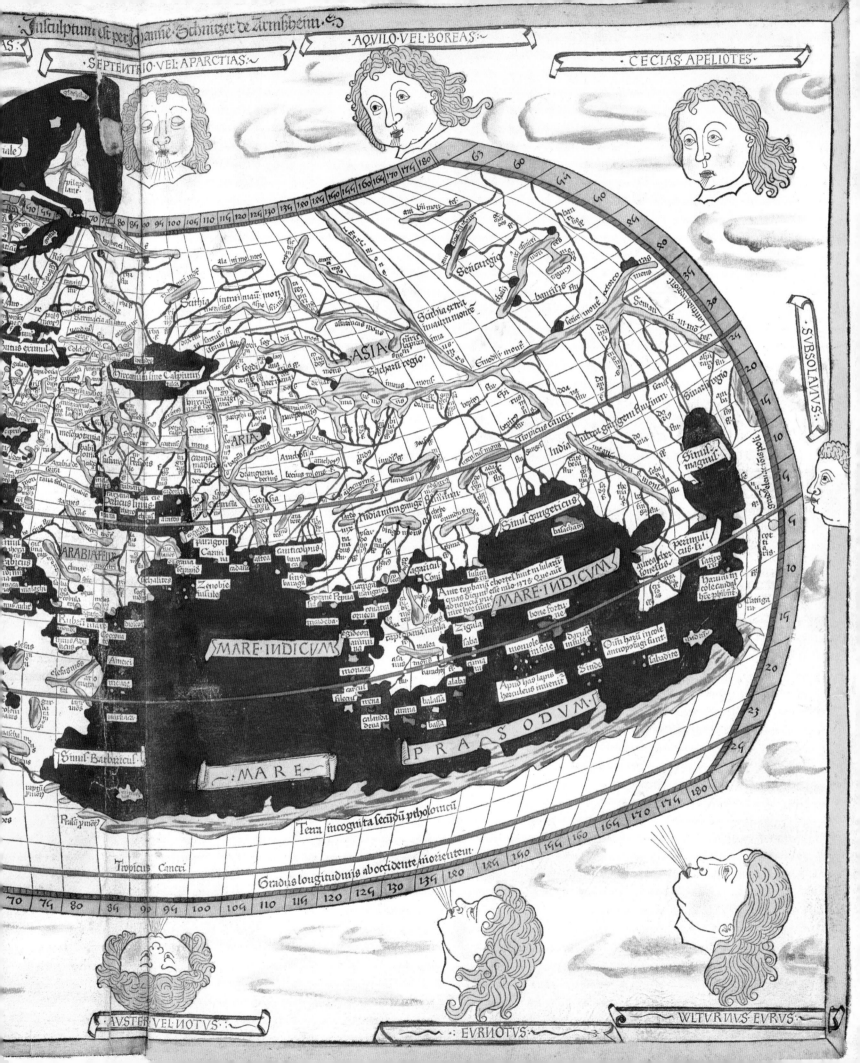

# The First Maps

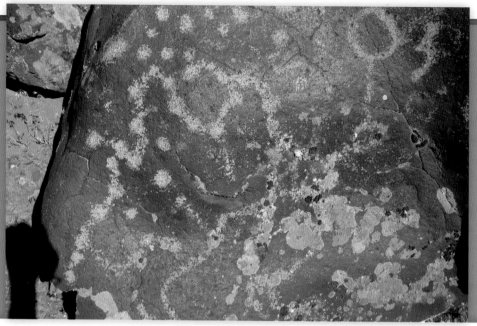

some 9,000 years ago. The 2.75-m (9-ft) wall painting shows what was once claimed to be an overhead plan of approximately 80 buildings, with an erupting volcano in the distance, which would have made it the oldest town plan in existence. However, in recent years a re-evaluation has led to the interpretation that the painting is far more likely to be of a leopard skin above a geometric design – that is, it might not be a map at all.

The oldest specimens that have been indisputably determined to be maps are those inscribed into clay tablets by the ancient peoples of Mesopotamia. Dating from about 2300BC, dozens of such map tablets have been found, including a detailed regional plan excavated from the ancient city of Nuzi (near Kirkuk, northern Iraq), which features rivers, cities and even a specific plot of land and who owned it. That plan is also the oldest known map to show orientation by the cardinal directions.

Another Mesopotamian tablet comprises the earliest known world map, produced around 600BC, as the empire of King Nebuchadnezzar II (reigned 604–562BC) neared the height of its powers. Not so literal a geographic interpretation as the Nuzi tablet, this map shows a flat Earth encircled by water, with seven triangular islands on the edges linking the Earth's oceans to a heavenly sea and the constellations. The Neo-Babylonians would have known there was more of the world that was not represented on this map, because although the centre of the map shows the Euphrates river and Babylon surrounded by eight other cities, it does not show Persia or Egypt, with which they were familiar. The map itself is quite small at just 125 × 75 mm (5 × 3 in), with

IT IS IMPOSSIBLE TO DETERMINE WHEN THE FIRST maps were created, not only because much of what was produced in prehistoric times has not survived or not yet been discovered, but also because it can be difficult to discern exactly what *was* devised as a map. For example, some archaeologists maintain that dot clusters in some of the images in the Great Hall of the Bulls at the Lascaux caves in France represent star maps of the main celestial constellations as they appeared to observers more than 16,000 years ago; other archaeologists have different theories, proposing instead that the dots symbolize hallucinations caused by sensory deprivation.

Ancient markings in other locations have been interpreted to be depictions of the routes of migratory animals or, in more arid regions, paths between waterholes. However, there is little consensus on this, as is shown by the archaeological debate over the origin of the numerous petroglyphs located along the Snake river in Idaho, USA, including the basalt boulder known as Map Rock. Dating from approximately 10,000BC, it is considered by some to be a map of the course of that river and its neighbour, the Salmon river. But others think this unlikely, and propose instead that it is an expression of the spiritual relationship between the native population, both animal and human, and the local land, which provided them with resources.

For a number of years, it was thought that one of the earliest surviving "maps" came from the Neolithic site of Çatalhöyük, in central Anatolia in modern-day Turkey, which was one of the world's largest settlements

Above left: Detail from a reconstruction of a Paleolithic rock painting in the Great Hall of the Bulls at the Lascaux caves in France.

Above right: There are numerous petroglyphs around Snake river. This boulder – which some think depicts a map of water resources – is near the Owyhee river, a tributary to the Snake in southwestern Idaho.

Below: The earliest known map of the world, incised on a clay tablet in Mesopotamia around 600BC. Above the map is a descriptive cuneiform text.

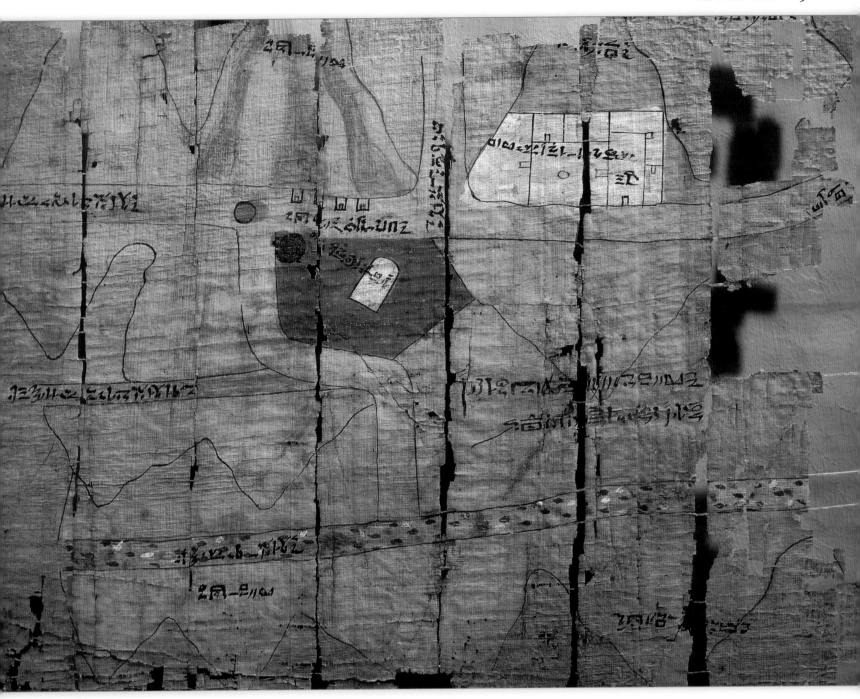

Above: The fragments of the Turin Papyrus, which in 1824 came into the possession of Bernardo Drovetti, an Italian serving as the French consul general in Egypt. Drovetti's large collection of Egyptian artefacts were the initial basis of the Egyptian Museum in Turin.

most of the tablet given over to explanatory text in cuneiform script.

The Egyptian civilization developed its maps independently of, but concurrently with, the peoples of Mesopotamia. Cosmological maps, designed to help the deceased through the afterlife, commonly decorated the inside of the coffin lid in royal tombs in the Egyptian period called "the New Kingdom" (1550–1070BC). However, other maps, some of which reveal the Egyptians' remarkable development of mathematics and surveying, are relatively rare, because they were produced on papyrus, a writing material derived from wetland sedge and quite fragile.

There remain a few ancient maps showing surveys of the River Nile, the extent of its floodwaters and the boundaries of estates in the most fertile areas of the region, but the most impressive surviving Egyptian map is the Turin Papyrus, taking its name from the Italian city where it is conserved. Painted on papyrus around 1150BC by the scribe Amennakhte for Pharaoh Ramesses IV (reigned c.1156–1150BC), the map shows part of the route to Wadi Hammamat, where blocks of *bekhen*-stone were quarried to produce large statues. Rendered in impressive detail, the Turin Papyrus shows prominent features of the region, including the long and winding *wadi*, the surrounding hills, a gold mine, the local human settlement and a quarry. It has been suggested that it is the world's first "geological map", because of its use of contrasting colours to represent different kinds of rock.

# Greek Mapping Traditions

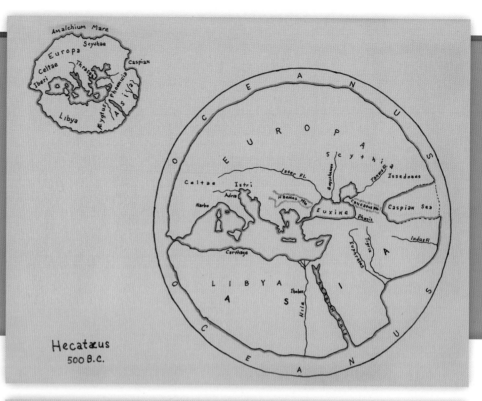

Hecatæus
500 B.C.

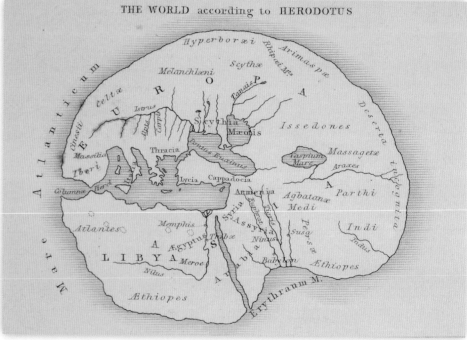

THE WORLD according to HERODOTUS

**I**N KEEPING WITH THEIR BRILLIANT WORK IN many intellectual spheres, the ancient Greeks of the Classical period (c.500–323BC) also laid the scientific foundations for the study of geography and the development of cartography. Sadly, few maps produced by the ancient Greeks have survived to the present, largely due to the fact that most were created using perishable media, such as papyrus or wood.

Miletus, an Ionian city on the western coast of Anatolia in modern-day Turkey, was home to two of the early great minds in cartography. Anaximander (c.610–c.546BC) is often considered the first Greek to have constructed a world map, in the sixth century BC, and about half a century later Hecataeus (c.550–c.490BC) produced the first known book on geography. This latter effort was expanded upon in the fifth century BC by the historian Herodotus, who used information gleaned from voyages, such as a supposed early circumnavigation of Africa by the Phoenicians, to delineate more clearly various parts of the world.

The idea that the Earth is spherical has been attributed to both Pythagoras (c.570–c.495BC) and Parmenides (c.515BC–unknown), but regardless of who originated it, by the mid-fourth century BC the proposal was widely accepted. By that time, another hypothesis of disputed parentage (Pythagoras, Parmenides or Aristotle) had divided the world into five climatic zones, with the centre zone hot, the zones either side of the centre temperate and the two at the ends of the Earth cold.

Around 300BC, Dicaearchus placed an east–west orientation line on a world map, and Eratosthenes, Hipparchus and Ptolemy thereafter successively developed the concept of dividing the globe into an imaginary grid of lines of latitude and longitude. In the third century BC, Eratosthenes (c.276–c.194BC), who for many years was in charge of the library at Alexandria, used geometry and the measurements of angles of shadows in two cities to

make the first known attempt to calculate the Earth's circumference. Sadly, the complete texts of both his great works – *On the Measurement of the Earth and Geographica (Geography)* – have been lost, and only parts of them remain in later summaries. This means there is no certainty to his exact measurements.

Hipparchus, an astronomer and mathematician of the second century BC, has been credited with dividing the circle (and therefore the Earth) into 360 degrees, a concept based on the ancient Sumerian sexagesimal numerical system, which had been passed down to the Babylonians. The sexagesimal model, with 60 as its base,

Top: The early Greek geographer Hecataeus produced this map of the world around 500BC. Note that it includes Libya as part of Asia, rather than as a separate continent.

Above: Herodotus's map of the world, as reproduced in James Rennell's *The Geographical System of Herodotus*, published in 1800. Rennell compared the theories of Herodotus to those of other ancient scholars, as well as to current geographical understanding.

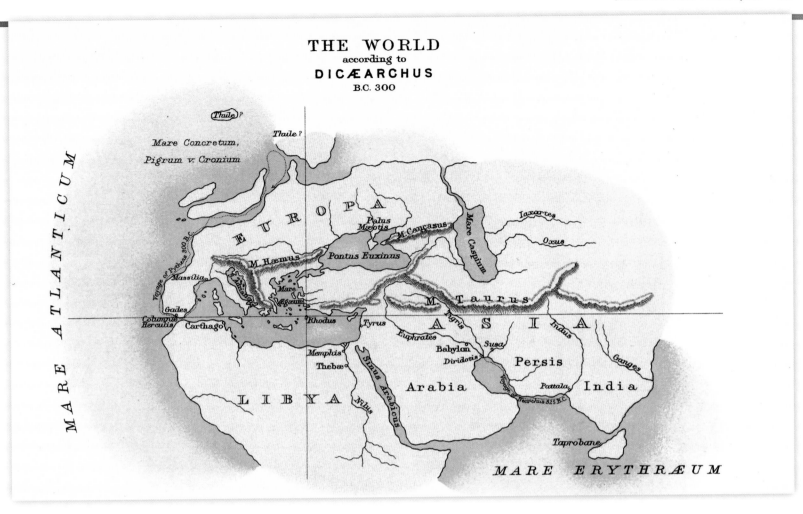

THE WORLD
according to
**DICÆARCHUS**
B.C. 300

*Mare Concretum,*
*Pigrum v. Cronium*

MARE ATLANTICUM

EUROPA

ASIA

LIBYA

Arabia

Persis

India

MARE ERYTHRÆUM

Above: The world map of
Dicaearchus, constructed around
300 BC, showed greater accuracy
in the coastlines of Europe and
Arabia, and included the first
east–west orientation line.

meant the subdivision of each of the 360 degrees into 60 minutes, and each minute into 60 seconds, which proved to be the crucial factor in the eventual development of geographic coordinates.

While scientists, mathematicians and theoreticians were key contributors to the development of Greek cartography, valuable parts were also played by travellers. An important part of the Greek understanding of geography came from *periploi* (singular *periplus*), the navigational manuscripts produced for centuries that listed ports, coastal landmarks, tidal conditions and other information of interest to sailors (and geographers). One famous example is that of Hanno, who sailed down the coast of West Africa in around 500 BC. Although there is little consensus as to whether Hanno actually reached Senegal, Sierra Leone or even as far as Gabon, his description of the "torrid region" seemed to confirm the existence of different climatic zones. Similarly, around 330 BC, Pytheas of Massilia sailed around Great Britain, and possibly as far as the Orkney Islands, the Shetland Islands, or even Norway or Iceland. This was the first recorded voyage to the far north by a Mediterranean navigator.

Perhaps the most famous of these navigational texts was the *Periplus Maris Erythraei* (*Periplus of the Erythraean Sea*), which gave first-hand details of the Red Sea, Indian Ocean and Persian Gulf, including ports in Egypt, Somalia,

Above: Several geographical theories
have been attributed to Parmenides,
but he is usually remembered as one
of the most significant pre-Socratic
philosophers.

the Arabian Peninsula and India. It is thought that this periplus was produced in the first century AD, because it is not mentioned directly in Strabo's *Geographica* (*Geography*). Completed no later than AD 24, when Strabo died, the 17-volume *Geographica* presented a descriptive account of places and people from around the known world. The massive work not only made use of *periploi*, it also provided an assessment of earlier geographers and cartographers, thereby recording for posterity the contributions of such men as Eratosthenes.

The greatest figure in the development of cartography in the ancient world is Claudius Ptolemaeus (*c.*AD 90–*c.*168), better known as Ptolemy. Although he spent much of his life in Alexandria under Roman rule, Ptolemy was a product of Greek civilization and thought, and he was able to synthesize Greek theory and Roman surveying practices to establish a foundation for mathematical cartography. Ptolemy's most significant work was the eight-volume treatise *Geographia* (*Geography*), which not only served as a gazetteer and physical description of some 8,000 locations, but also gave coordinates, which allowed the places to be plotted on maps. His great work also included an in-depth discussion of map projections and detailed instructions for mathematical map-making at both the global and regional levels. Although no maps from Ptolemy's masterpiece have survived, his cartographic ideas have influenced map-making ever since.

# Roman Maps and Itineraries

Below: A fragment of the Forma Urbis Romae, showing part of the Subura, a notorious area of Rome that included a red-light district. The main street in the section is the Clivus Suburanus.

THE SIGNIFICANT PHILOSOPHICAL AND CULTURAL DIFFERENCES between the Greek and Roman civilizations were sometimes reflected in maps. The Romans cared less than the Greeks did about the theoretical and mathematical aspects of cartography, and instead they viewed map-making from a practical, organizational standpoint. To Roman leaders, geographical surveying and recording were simply tools with which to further their military, political and administrative ends.

An example of this cartographic dissimilarity is provided by Julius Caesar's commission of a geographical survey of the Roman world in 44BC. The overseeing of this task eventually fell to Marcus Vipsanius Agrippa, a close ally of and advisor to the young Octavian (later Caesar Augustus), who continued the survey when Agrippa died in 12BC, before the project could be completed. Within a few years the map was engraved in the marble of the monument called the Porticus Vipsania to emphasize the power and glory of the empire. Although the structure has not survived, the map can be reconstructed to a great extent from *Naturalis Historia* (*Natural History*) by Pliny the Elder (AD23–79).

Around AD43 Pomponius Mela produced *De Situ Orbis* (*A Description of the World*), which is the only ancient treatise on geography in classical Latin. The work shows the influence of Parmenides and other Greeks, in that Pomponius Mela divided the world into five climatic zones. He noted that both temperate zones were habitable, but although he stated that the southern zone of the two was lived in by a people called the Antichthones, its geography was unknown

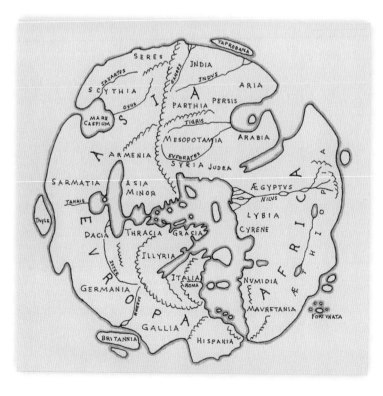

Opposite below: A reconstruction of the world map compiled under the direction of Marcus Vipsanius Agrippa. Anyone leaving Rome by way of the Via Flaminia would see from this map on the Porticus Vipsania what a major part of the world was controlled by Rome.

Below: The section of the Madaba Mosaic showing a layout of the city of Jerusalem. At the very centre of the photo is the Church of the Holy Sepulchre.

because of the heat of the intervening zone. This work remained influential for more than 1,300 years.

The Romans were less interested in world geography than they were in surveying local lands and in carefully mapping estates and cities. One of the most remarkable examples of this is the first known detailed map of Rome, called the Forma Urbis Romae (or Severan Marble Plan of Rome). Produced in marble between AD203 and 211, the enormous final map measured approximately 18 × 13 m (60 × 43 ft) and covered a wall inside Emperor Vespasian's Templum Pacis (Temple of Peace). The map included every architectural feature in Rome. Today, only 1,186 fragments survive, accounting for 10–15 per cent of the original.

The Roman Empire had a road system that extended for more than 80,000 km (50,000 miles) throughout Europe, North Africa and parts of Asia. For military, administrative, postal and general travel, more common than maps were itineraries, which were written lists of places along any particular road. (In fact, maps have been described as "painted itineraries".) The most famous one still in existence is the Antonine Itinerary, which was most likely produced in the third or fourth century AD, and provides several thousand place names, with the distances between settlements or other significant places to stop.

The best example of what an actual Roman road *map* would have looked like is the document known as the Peutinger Table. The original is believed to have been drawn in the fourth century, but the Peutinger Table is a copy made more than 800 years later that was only discovered after several centuries and given in 1508 to the German antiquarian Konrad Peutinger, after whom

it was named. The map is in 11 sheets of parchment approximately 7 m (23 ft) in length, but only 30 cm (1 ft) in width, perhaps because the original was on a long papyrus roll. Due to its unusual shape, the map's scale is greatly compressed in a north–south direction, whereas it extends lavishly all the way from the western edges of the Roman Empire to India in the east. Among the more than 4,000 features represented are roads, cities, villas, temples, granaries, harbours and lighthouses, rivers, forests and mountains. Because of its importance, the Peutinger Table was placed on the UNESCO Memory of the World Register in 2007.

By the fifth century, the Western Roman Empire was in slow and terminal decline, but the Eastern Roman Empire continued to flourish. Perhaps the most remarkable map surviving from the Byzantine East is a vast, mid-sixth-century mosaic in the Greek Orthodox parish church of St George in Madaba in modern-day Jordan. The Madaba Mosaic was discovered in the 1890s when the church was being built over the remains of a Byzantine church. Originally, the mosaic would have covered virtually the entire floor area of the former church, with a size estimated variously at from 15.7 × 5.6 m (52 × 18 ft) up to 24 × 7 m (79 × 23 ft). About 25 sq m (270 sq ft) remain of this incredible map of the Holy Land, which extends from the Mediterranean to east of the River Jordan, and from Syria south to the Nile Delta. Cities, towns, oases, lakes, rivers, bridges and key religious sites are all colourfully depicted. The mosaic is so detailed and accurate that what survives has been used to verify the location of previously unknown biblical sites.

14

Below: Two sections of a 1753 edition of the Peutinger Table, as engraved by the German draughtsman and artist Salomon Kleiner. They show (below) a section of the Roman Empire that later became Austria-Hungary and (bottom) the lands to the east of the Mediterranean, including Syria, Arabia, Mesopotamia and India.

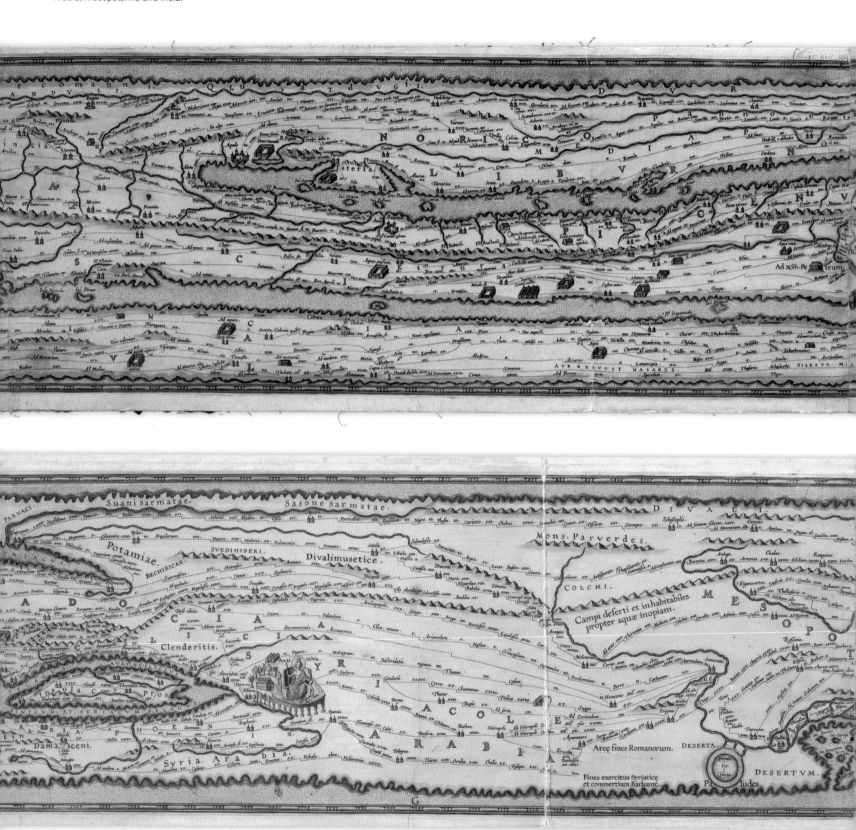

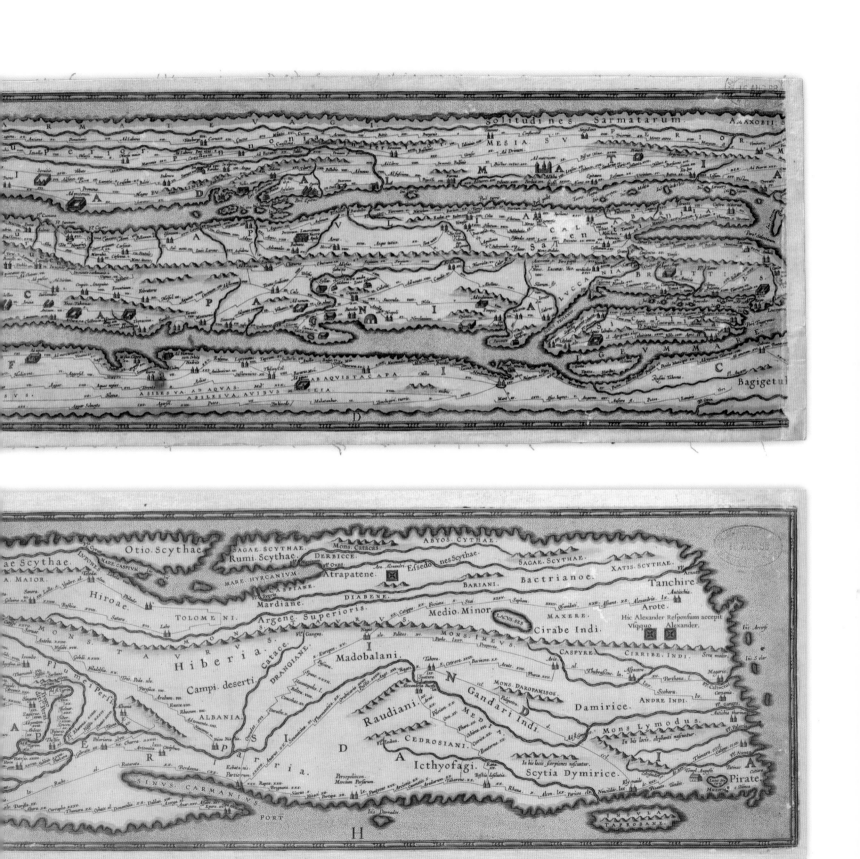

# Chinese and Pacific Society Maps

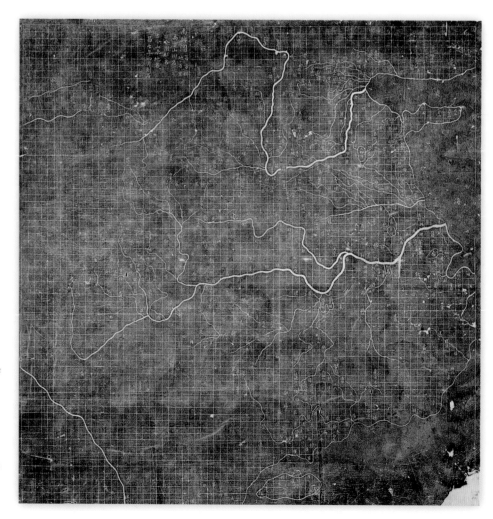

Above: Silk map from the Han Dynasty in the second century BC, found in a tomb in Mawangdui in Hunan. The map is oriented with south to the top.

Below: The Yu Ji Tu includes impressive detail of the river systems of China. It is overlaid by a grid of squares, giving it a precise scale.

A S WAS THE CASE WITH SO MANY OTHER artistic and scientific endeavours, cartography in China developed independently from that of the civilizations in the Near East, North Africa and Europe. Cartography also arose separately in other regions of East Asia, as well as in the islands of the South Pacific.

Although Chinese maps were undoubtedly produced earlier than the third century BC, the oldest extant ones date from that time. These were discovered in 1986 in a pre-imperial Qin kingdom tomb in Gansu province. The maps were drawn in black ink on pieces of wood that measured about 27 cm (10.5 in) in length and up to 18 cm (7 in) in width. The wood needed to be slow-dried for two years in a restoration process before it could be determined by archaeologists that the maps show three river systems, as well as local roads, settlements and mountains.

Previously, the oldest known Chinese maps had been three produced in around 168BC and unearthed in 1973 from a Western Han period (206BC–AD23) tomb in Hunan province. These maps – one topographic, one military and one of a city – were drawn on silk and show an area in south-central Hunan. The three cover a larger area than the Qin maps, with more-detailed information and numerous well-developed map symbols, and they are thought to be the earliest surviving maps in the world to have been compiled on the basis of field surveys.

A major development in Chinese mapping occurred in the third century AD, when Pei Xiu (AD224–271), a geographer, cartographer and government administrator, established his "six principles" – emphasizing scale, location reference, distance, elevation, direction and gradient – that would guide Chinese map-making for the next 1,400 years. The great weakness of Pei Xiu's work was that he did not address issues such as map projection and spherical coordinates, in great part because Chinese maps did not cover a broad

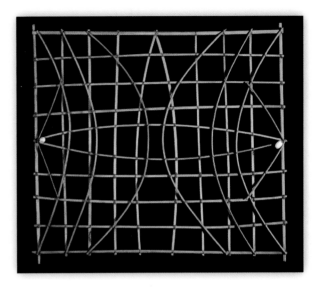

Left: The Da Ming Hun Yi Tu (Great Ming Amalgamated Map), produced about 1389, showing the numerous paper labels in the Manchu language superimposed on the Chinese original place-names.

Right: A stick chart from the Marshall Islands. This small version is a *mattang*, of the kind used to teach about wave and swell patterns around one island.

Below: The Gotenjiku Zu, or Map of the Five Indias, was drawn by the Japanese Buddhist monk Juaki in 1364. It is based on a Chinese priest's account of his journey to India in the seventh century and even includes a red line tracing the pilgrimage route.

enough region to make the Earth's curvature a major cartographic factor. These problems would not be resolved for many centuries.

In the sixth century, Chinese cartographers began using stone as a medium, carving intricate maps on stelae. One of the most remarkable stelae is the Yu Ji Tu, or Map of the Tracks of Yu, which was created in 1137. Roughly 90 cm (3 ft) square, it contains a grid system of squares overlaying the map, with each square representing 100 *li*, or about 53 km (33 miles). The Yu Ji Tu shows China's coastline and river systems, and lists settlements, lakes and mountains; it clearly served as an important tool for those ruling a vast area. This kind of map was the earliest example of one from which multiple copies could be generated without having to produce them by hand – because the use of stone made it possible for ink rubbings to be transferred onto paper (which had been invented by the Chinese in the second century AD).

Conquest by the Mongols in the thirteenth century linked China with other subjugated territories extending thousands of miles to the west, which encouraged Chinese map-makers to include new data about outside regions of the world. The spread eastward of Islamic maps had a similar effect. The oldest surviving Chinese world map is known as the Da Ming Hun Yi Tu (Great Ming Amalgamated Map), and it was produced in about 1389 from an older map – in order, it has been suggested, to celebrate the triumph of the native Chinese Ming dynasty over the Mongol Yuan dynasty. Painted on silk at a size of 3.9 × 4.6 m (12 × 15 ft), the map placed China at the centre of the world, which was shown to extend from Japan in the east to Europe in the west and from Mongolia in the north to Java in the south. The written characters of the original Chinese were covered by labels in Manchu when the Manchurian Qing dynasty came to power in the seventeenth century.

Although cartography in Japan and Korea evolved separately from China at both a regional and a local level, map-makers in both countries were greatly influenced by the Chinese in the development of "world" maps.

Among the cultures of the South Pacific, totally different kinds of map developed. At a time when people did not have advanced mechanical aids to navigation, nautical information was maintained first by mental map images formed from aspects of the natural environment. These included the stars, prevailing winds and movement of weather patterns, all of which helped mariners establish their location. However, they were also aided by extremely complicated devices known as "stick charts". Made from thin strips of coconut frond or pandanus root, the "sticks" were arranged in patterns that depicted the direction and strength of currents, swells and surface waves, with cowrie shells or coral pebbles used to represent islands or atolls.

Many peoples of the South Pacific – including those of the Marshall Islands, Caroline Islands, Gilbert Islands, Solomon Islands and Cook Islands – used these stick charts, which were generally broken down into three kinds. The *mattang* was a small chart that showed wave patterns around a single island and was used for teaching. A *meddo* was a larger chart that showed a limited number of islands and the sea movements between them, as well as influenced by them. And a *rebbelib* was a more general navigational aid for a major island group or groups. Unlike traditional maps, stick charts were consulted prior to going to sea, but they were not taken on voyages; the wave and swell patterns all had to be gauged by experience.

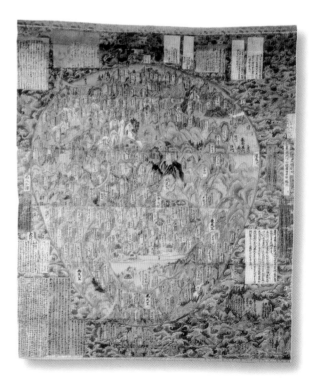

# Islamic Maps
## and Cartography

Right: A page from *Survey of Roads and Kingdoms* by Abu Ishaq Ibrahim Muhammad al-Farisi al-Istakhri, illustrating the simple but highly stylized representations of the Balkhi school.

FOLLOWING THE DEATH OF MUHAMMAD IN AD632, the religion of Islam spread from the Arabian Peninsula as far as Central Asia and the Iberian Peninsula. With such a great expansion in the geographical boundaries of the Islamic world, it was natural that this should be accompanied by a growth of interest in mapping that world, and by the tenth century there had developed two distinct Islamic mapping traditions. The first was closely related to the aesthetics, traditions and religious tenets of the Islamic world. Founded in the early tenth century by Abu Zayd Ahmad ibn Sahl al-Balkhi (*c.*850–934), what has become known as "the Balkhi school" produced geographies and maps that reflected the political and religious views of Islamic leaders. The second incorporated many of the traditional theories of Ptolemy, whose *Geographia* had been translated into Arabic in the ninth century, along with many other ancient Greek documents about astronomy, mathematics and cartography.

The format of al-Balkhi's classic work *Suwar al-Aqalim* (*Picture of the Climates*) was followed by several of the most famous Islamic cartographers: Abu Ishaq Ibrahim ibn Muhammad al-Farisi al-Istakhri in his *Kitab al-Masalik wa-al-Mamalik* (*Survey of Roads and Kingdoms*) around 950, and Abu al-Qasim Muhammad ibn Hawqal in his treatise *Kitab surat al-Arb* (*Picture of the Earth*) in 1086. Each of the maps in these collections was highly stylized, tending to use only vertical and horizontal lines, 90° angles and circles or arcs for their primary features. Each also showed different features in specific colours, such as mountains in red and water in blue, other than the River Nile, which was also coloured red. The style of these maps has been likened to the familiar depiction of the London Underground.

The works of the Balkhi school tended to include 21 maps, one of the world, one each of the Mediterranean Sea, the Indian Ocean and the Caspian Sea, and other regional charts. These incorporated information about settlements, trade and pilgrimage routes, and oases. The Balkhi school maps were more concerned with political boundaries than geographical detail, and each one was centred on Mecca, indicated by a symbol of the Kaaba, the cube-shaped building that is the most sacred site in Islam.

However, many historians of cartography believe that Islamic cartography reached its apex with the work of Abu Abdallah Muhammad ibn Muhammad al-Sharif al-Idrisi (*c.*1099–*c.*1160), a Moroccan geographer who was heavily influenced by Ptolemy and who lived much of his life at the court of King Roger II, the Norman ruler of Sicily. It took al-Idrisi 15 years to complete his masterpiece, which appeared in 1154 as *Nuzhat al-Mushtaq Fi'khtiraq al-Afaq* (*The Book of Pleasant Journeys into Faraway Lands*), but is frequently called the *Al-Kitab al-Rujari* (*The Book of Roger*). The book was essentially an atlas with a world map and 70 regional maps, each complete with detailed physical, political and cultural information about the areas depicted. Following the Ptolemaic system, the inhabited world was divided into climate zones, in this case seven from north to south – each of which had 10 regions from east to west, producing the total of 70 regional maps. Al-Idrisi believed the known world to cover only 160° of longitude; each regional map therefore covered 16° from east to west.

Despite their underlying differences, there are notable similarities between the maps of the Balkhi school and al-Idrisi. All tend to be oriented with south at the top, making them slightly ungainly to Western eyes. Moreover, even the ones influenced by Ptolemy followed the Islamic aesthetic of using bright colours and flowing Arabic script. On many of al-Idrisi's maps, the rivers are in green, the seas in blue and the mountains in purple and ochre.

# Advances in Surveying Instruments

The beautiful aesthetics of the Islamic mapping tradition can obscure the fact that the survey instruments and techniques used in the Islamic world were the most advanced of the time. Centuries earlier, the Romans had used two primary instruments for ground surveys: the *groma* and the *chorobates*. The *groma* had originated in Mesopotamia and been imported by the Greeks; it was used to survey straight lines and right angles, and consisted of a vertical iron staff with horizontal arms just over 1 m long (3.5 ft) that were mounted at right angles on a bracket. Each end of the cross arms had a plumb line hanging vertically. The *chorobates* was a bar just over 6 m (20 ft) long with perpendicular legs and diagonal pieces between the bar and legs. The diagonals had vertical lines, over which plumb lines were hung. When the instrument was level, the plumb lines struck both vertical lines. By using two level instruments, the vertical distance between them could be determined by sighting along the depth of the higher, to a rod placed atop the lower.

By expanding on the Greek tradition, the Islamic world refined old instruments and invented many new ones that could be used for navigation, astronomy and cartography in a much broader context than the Roman ones. For example, the astrolabe can be traced back to Classical Greece, but it was developed to a higher level when Islamic astronomers introduced angular scales and added circles to indicate azimuths on the horizon. Similarly, in eighth-century Persia, astronomers produced an improved version of the Greek armillary sphere, a device used to demonstrate the movement of celestial bodies around the Earth or the Sun. The spherical astrolabe, a variation of the astrolabe and the armillary sphere, was invented in the Islamic world in the ninth century. In the next century, the mural sextant was developed in Persia. And at least four different kinds of quadrant were improved upon or invented by Islamic astronomers in Iraq, Syria and, possibly, Egypt.

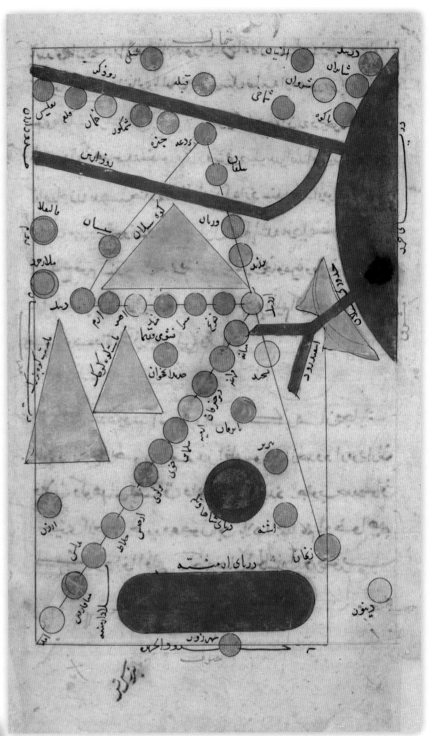

Left: An astrolabe originally produced in Italy around the beginning of the sixteenth century. It was used to determine latitude by measuring the altitude of the Sun or a star of known declination.

Above: A fourteenth-century Persian adaptation of a map of Armenia and Azerbaijan by al-Istakhri. The Caspian Sea is at the top right and the city of Ardabil is slightly right of centre, at the junction of three roads.

Below: The world map from *The Book of Roger*, produced in the mid-twelfth century by Abu Abdallah ibn Muhammad al-Sharif al-Idrisi for Roger II, the Norman king of Sicily. It is oriented with south to the top.

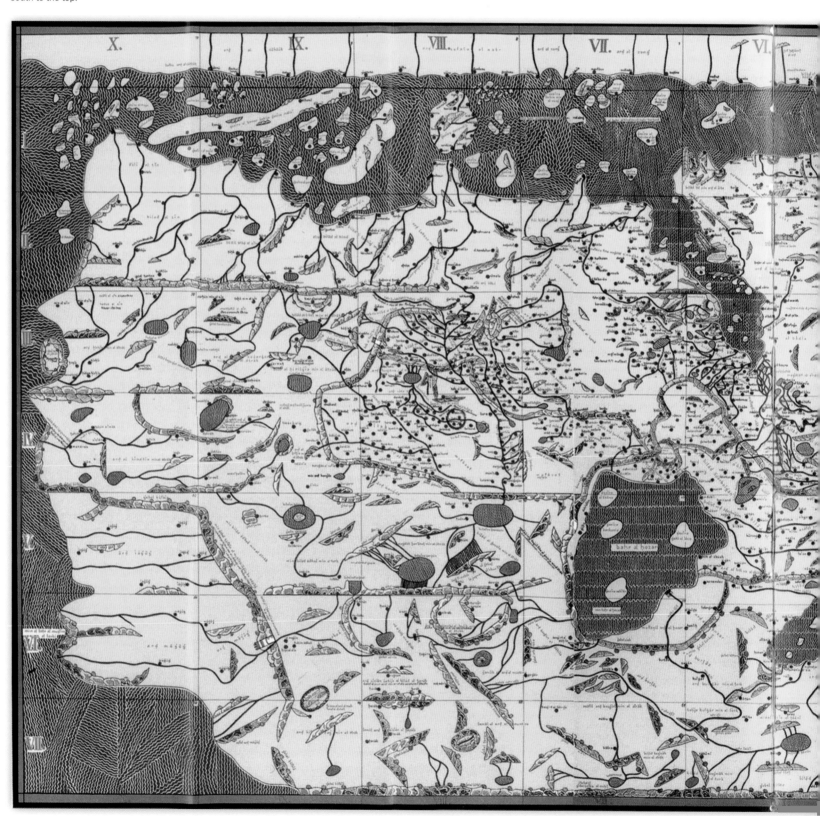

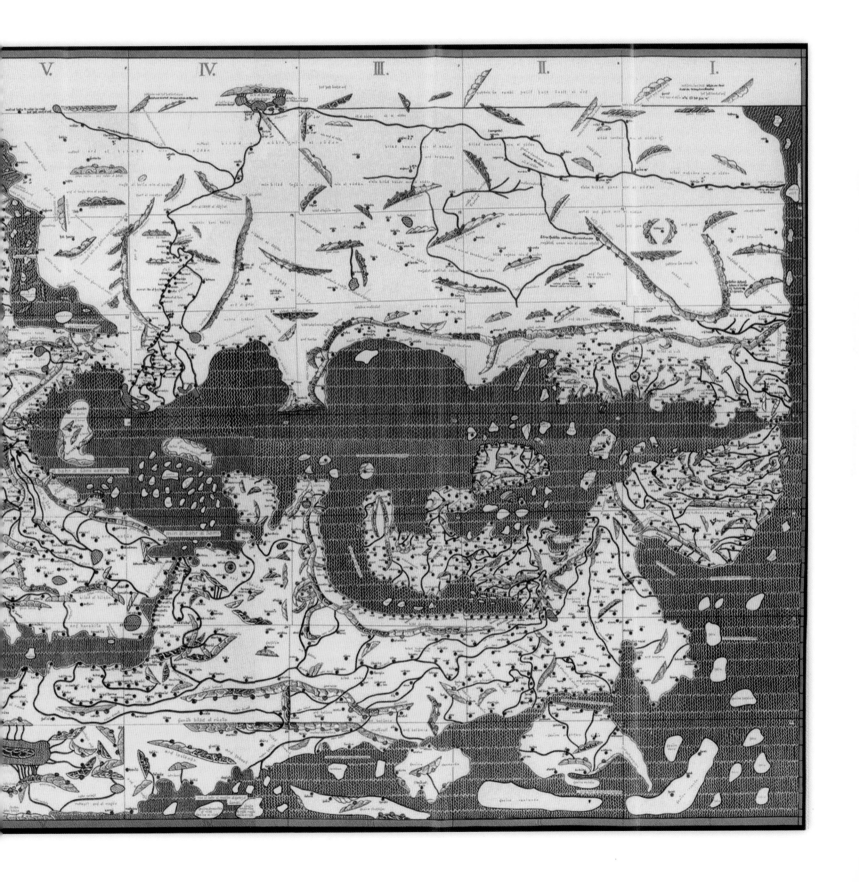

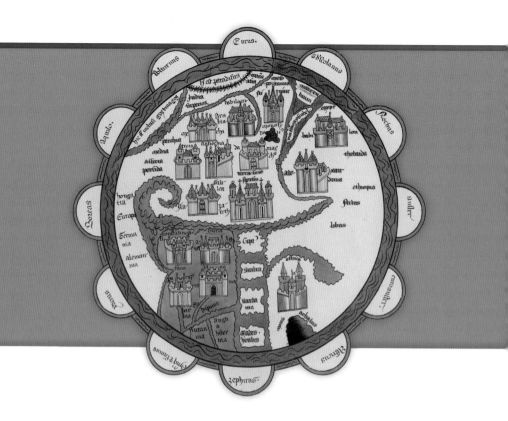

# Medieval Maps and Charts

To understand medieval European maps – particularly the renowned *mappae mundi*, the "maps of the world" (taken from the Latin *mappa* for "cloth" and *mundi*, "of the world") – one must comprehend the mentality that pervaded Western thought in the early Middle Ages (*c*.410–*c*.1000). Daily existence was dominated by the Roman Catholic Church, and that affected cartography as much as any other aspect of life. Map-makers were not simply recorders of geographical or topographical features, but men obligated to reflect in their work the ecclesiastical doctrines handed down by the Church. Therefore, maps were not just projections of physical surveys, they also embodied the Christian world view of a divine order. In this sense, the essence was not dissimilar to that of the Islamic Balkhi school, in that the maps reflected religious views rather than strictly geographical ones.

The most common *mappae mundi* are the tripartite or "T-O maps", so called because they look like a T within a circle. The T – which also symbolically represented the cross (in its Greek Tau form) – divides the Earth into the three known continents (Asia, the largest, is usually at the top, with Europe to the bottom left and Africa to the bottom right). The vertical of the T represents the Mediterranean, separating Europe from Africa. The horizontal of the T represents the Black Sea and the River Don, dividing Europe from Asia, and the Red Sea and the River Nile, separating Africa from Asia. The O was often considered to represent the encircling ocean, similar to how the early Greeks envisioned the world, but it is certain that many cartographers knew the Earth to be spherical rather than disk-shaped, and that this was simply a schematic rather than a geographical feature. Asia is the largest area in T-O maps because it was, according to St Augustine, "the most blessed". In addition, Asia tends to be located at the top of the maps, not only because the Sun

Top: The world map from the Chronicles of St Denis, as reproduced in the Viconde de Santarem's 1840s *Atlas Compose de Mappemondes*. Rather than be taken literally, this map was intended to convey a conceptual view of the world.

Above: A reproduction in Santarem's *Atlas Compose de Mappemondes* of a world map originally created by Beatus of Liébana for his *Commentary on the Apocalypse*. At the top are Adam, Eve and the serpent in Paradise.

Opposite: A reproduction of one of the maps of Britain created by Matthew Paris. The rivers are over-emphasized and the shape of Scotland bears little resemblance to reality.

MAP A. COTTON MS. CLAUDIUS D. VI, FOL. 12ᵛ

rises in the east, but also due to the tradition that the Garden of Eden lay in the easternmost part of the world. (In fact, the word "orientate" comes from *oriens*, the Latin for "east".)

Another type of *mappae mundi* is the quadripartite or Beatus map, named for Beatus of Liébana (*c*.730–*c*.800), a Spanish Benedictine monk. Perhaps the best known of these is one that accompanied Beatus's popular *Commentaria in Apocalypsin* (*Commentary on the Apocalypse*), which he first produced in 776. A map was added to the text in order to define the world into which the 12 apostles were sent to proselytize. As with many such maps, the originals are no longer extant and much of the knowledge about them has come through the study of later reproductions. Some of those reproductions are roughly square with rounded corners, while others are more spherical, although both kinds still include the encircling ocean. Although orientated with the east at the top, the maps have one major difference from the T-O maps: to the right – the extreme south – is a fourth continent separated from Europe and Africa by a red waterway and marked as where the "Antipodeans" lived.

Roman-style itineraries continued to be used instead of picture maps into the early Middle Ages. The best known of these was compiled in around 700 by a monk from Ravenna on the Adriatic coast. Known as the *Ravennatis Anonymi Cosmographia*, or *Ravenna Cosmography*, it lists more than 5,000 place names, including kingdoms, islands, native peoples, rivers and seas as far afield as Ireland and India. Many of the names were taken straight from Roman sources, including all those from Britain.

Even one of the greatest of all medieval map-makers still used itineraries as a source. Matthew Paris (*c*.1200–1259) was an English Benedictine monk, who spent most of his adult life in an abbey at St Albans, where, between about 1240 and 1253, he produced a series of

histories, which includes a three-volume masterpiece known as the *Chronica Majora* (*Major Chronicles*). Fifteen of the maps created to accompany his texts still survive, including two of Palestine, four multi-page strip maps of an itinerary from London to Rome, and four versions of a map of Britain. The maps of Britain are exceptional, in that no other national or regional maps of Europe are known to have been produced during that period. They are also unusual in that they are oriented with north to the top.

Paris appears to have taken the data for his maps from a variety of sources. The coastal outline of Britain seems to be from a world map, and he used an already-existing itinerary for the names and locations of rivers, settlements and roads. He then added additional information from his own studies, accounts from travellers and contemporary writing to fill in more names of towns, castles, waterways and notable geographic features.

Paris included some roads in Britain, but the oldest surviving map of Britain with a true emphasis on its roadways was produced about a century later. Generally dated at around 1360, it is known as the Gough Map, for the eighteenth-century antiquarian Richard Gough, who donated it to the Bodleian Library at Oxford. The map was produced on two joined pieces of vellum, measuring approximately 115 × 56 cm (45 × 22 in), and was drawn in ink with coloured washes.

Oriented with the east at the top, the Gough Map's geographic

Below: The Gough map, which is oriented with east to the top. All of the major towns of the time are included, but neither Scotland nor the smaller surrounding islands resemble their true forms.

Opposite: A section of Battista Agnese's *Portolan Atlas* of 1553. This map includes the Atlantic Ocean and its surrounding regions, which comprised much of the known world, but does not show Scandinavia or the Far East.

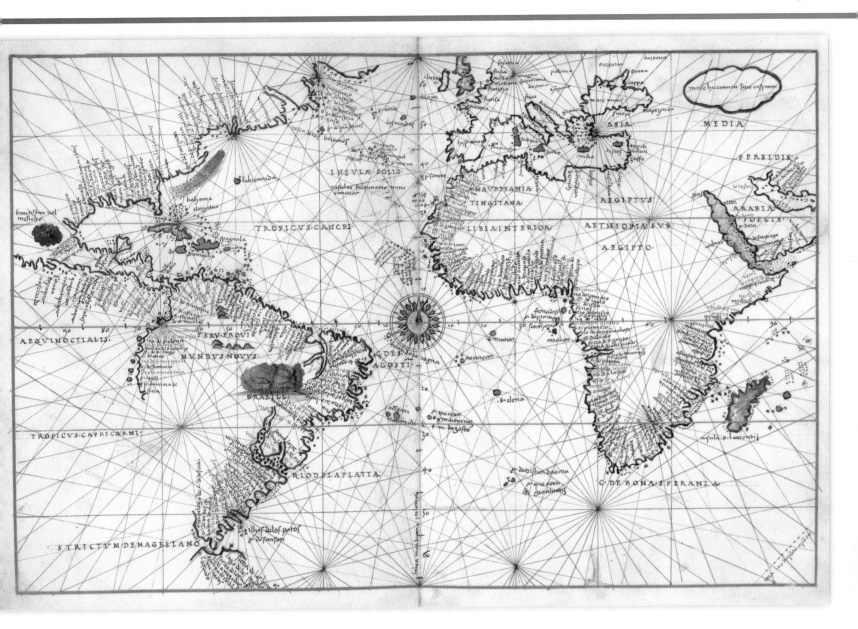

outline of England is excellent, although Scotland is less so. Rivers such as the Thames and Severn receive great emphasis, and many towns are shown with lavish illustration – particularly London and York, for which the lettering is coloured gold. However, there is little written commentary, unlike many other maps of the time. Various other features are identified with symbols, such as a tree for the New Forest and the numerous castles in Wales constructed on the orders of King Edward I. Routes between towns are marked in red, with Roman numerals indicating distances measured in leagues.

In the period between the works of Paris and the unknown creator of the Gough Map, remarkable advancements were made in a new area of cartography. For thousands of years, European mariners had tended to stay within sight of land, and because of this no great effort had been made to develop true navigational charts. But in the twelfth century, the magnetic compass, which had been developed in China centuries earlier, was first widely introduced in Europe. When the compass was combined with other basic navigational aids, such as a log line, it became possible to sail directly from one port to another rather than hugging the coast.

A trade at sea soon blossomed, particularly from city-states such as Venice, and by 1270 there are records of maritime charts being used. These are known as portolan charts – from the Italian word *portolano* ("related to ports or harbours") – or pilot books, which listed ports, anchorages and sailing instructions in much the same way as the itineraries had for land. Produced in maritime centres such as Genoa, Venice, Barcelona and Majorca, portolan charts were initially hand-drawn on vellum or another form of animal skin. The charts had certain consistent features, which included being criss-crossed by rhumb lines (lines that cross all the meridians of longitude at the same angle) radiating from a series of compass stars. Each compass star was the centre of 16 or 32 rhumb lines. Eventually, compass points and rhumb lines were coded in colour to aid navigators.

Portolan charts also featured a scale, and because they were intended to be used in changing sea conditions, they were not orientated with a top or bottom, but designed to be rotated as desired. In many charts, detailed coastlines were included, although some features were deliberately distorted, with certain capes, estuaries and harbours enlarged to provide the navigator with information about key

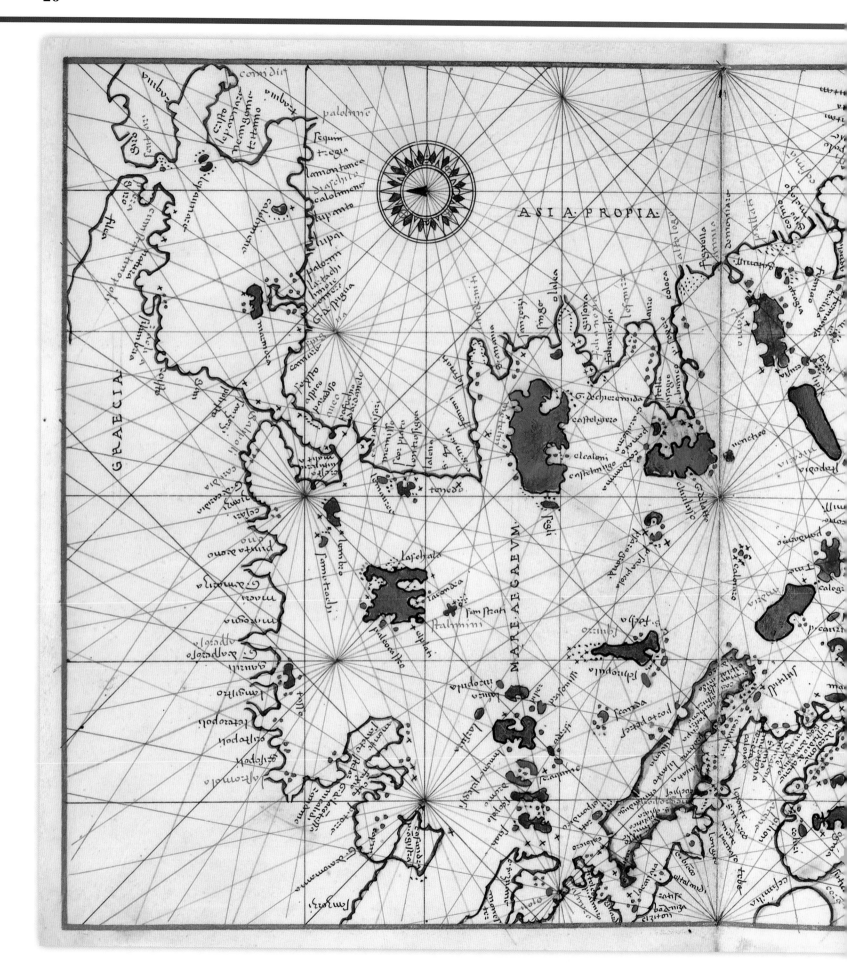

ASIA·PROPIA·

GRAECIA·

MARE·AEGAEVM·

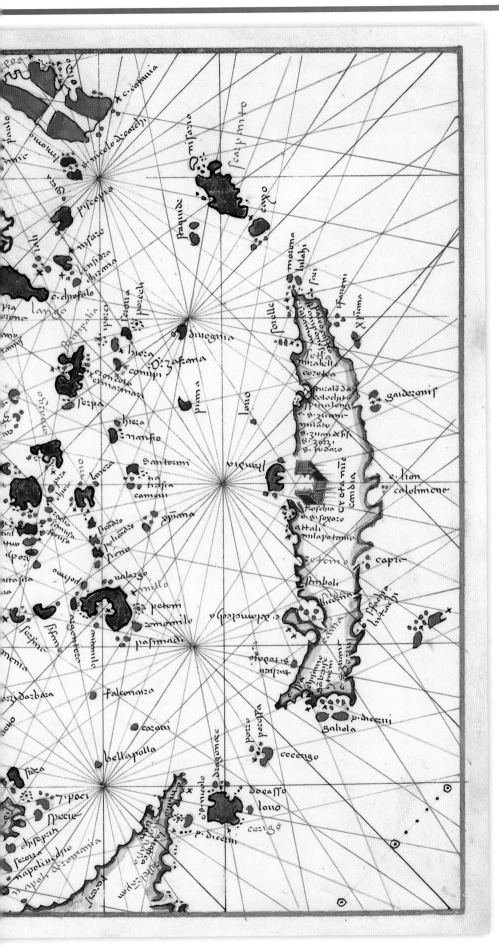

approaches, hazards or sources of supplies. As portolans were revised and refined, they became more reliable, although one mistake was repeated many times – namely, that a degree of longitude was equal to a degree of latitude. This error caused little significant distortion in the Mediterranean Sea, but it proved problematic in maps created for higher latitudes.

The oldest surviving portolan chart dates from the late thirteenth century and is known as the Carte Pisana, having once been owned by a wealthy family from Pisa. Hand-drawn on sheepskin, the chart depicts the entire Mediterranean, and it contains most of the elements that would become commonplace in later portolan charts.

Left: A section of Battista Agnese's *Portolan Atlas* of 1553, which was published in Venice and bound in Moroccan leather. This map, which is oriented to the west is of the Aegean Sea and its surrounding lands.

Below: An elaborate compass rose from Charles Massey's mid-seventeenth-century portolan chart A Draft of the Bay of Coche and R. Nagor on the Coast of Guzzaratt.

It was a logical progression from portolan charts to atlases. However, a distinct difference developed between those atlases created in Genoa or Venice and those originating in Catalonia on the Iberian Peninsula. Whereas the former concentrated on their navigational function, the latter were heavily influenced by the aesthetics of Islamic art and cartography, due to the proximity of the Islamic domains. The finest such work is known as the Catalan Atlas, and is attributed to Abraham Cresques, a Jewish "book illuminator" from Palma in Majorca. Originally prepared around 1375 at the behest of King Pedro IV of Aragon, the atlas consisted of six leaves of vellum stretched over wood panels, which were later divided into a dozen half-sheets. The first four half-sheets comprise discourses on cosmography, astronomy and astrology, with illustrations that show a strong Islamic influence. They also include the earliest known set of lunar tide tables and an astrological wheel with pictures of the signs of the zodiac.

The remaining sheets of the Catalan Atlas form the world map, which conforms to one of the dictates of Islamic cartography by being oriented with south at the top. With a great deal more information available for Europe and the Mediterranean region, the cities and geographical features of those areas are more accurately and more intensively plotted. By contrast, corresponding parts of Asia and Africa are considerably more adorned with artwork to offset the paucity of information. The atlas also contains perhaps the first example of a compass rose, a highly ornate figure at the intersection of rhumb lines, showing the directions and bisections of the winds as on the older wind rose, and including a pole star that indicates north. Cresques also followed European tradition in identifying Christian cities with a spire and cross, while he initiated the custom of indicating major Islamic settlements with a dome.

Right: A map of the Europe and the Mediterranean world taken from Abraham Cresques's Catalan Atlas. The two pieces of the map overlap, causing some features, such as Sardinia and Corsica, to be drawn twice.

Overleaf: Two half-sheets from the Catalan Atlas of Abraham Cresques. The map is oriented with south at the top, although some figures appear upside-down regardless of which way it is viewed. This version was produced by the Visconde de Santarém in his nineteenth-century atlas.

*Carte Catalane de 1375,* [illegible] *à la Bibliothèque Nationale de Paris.*

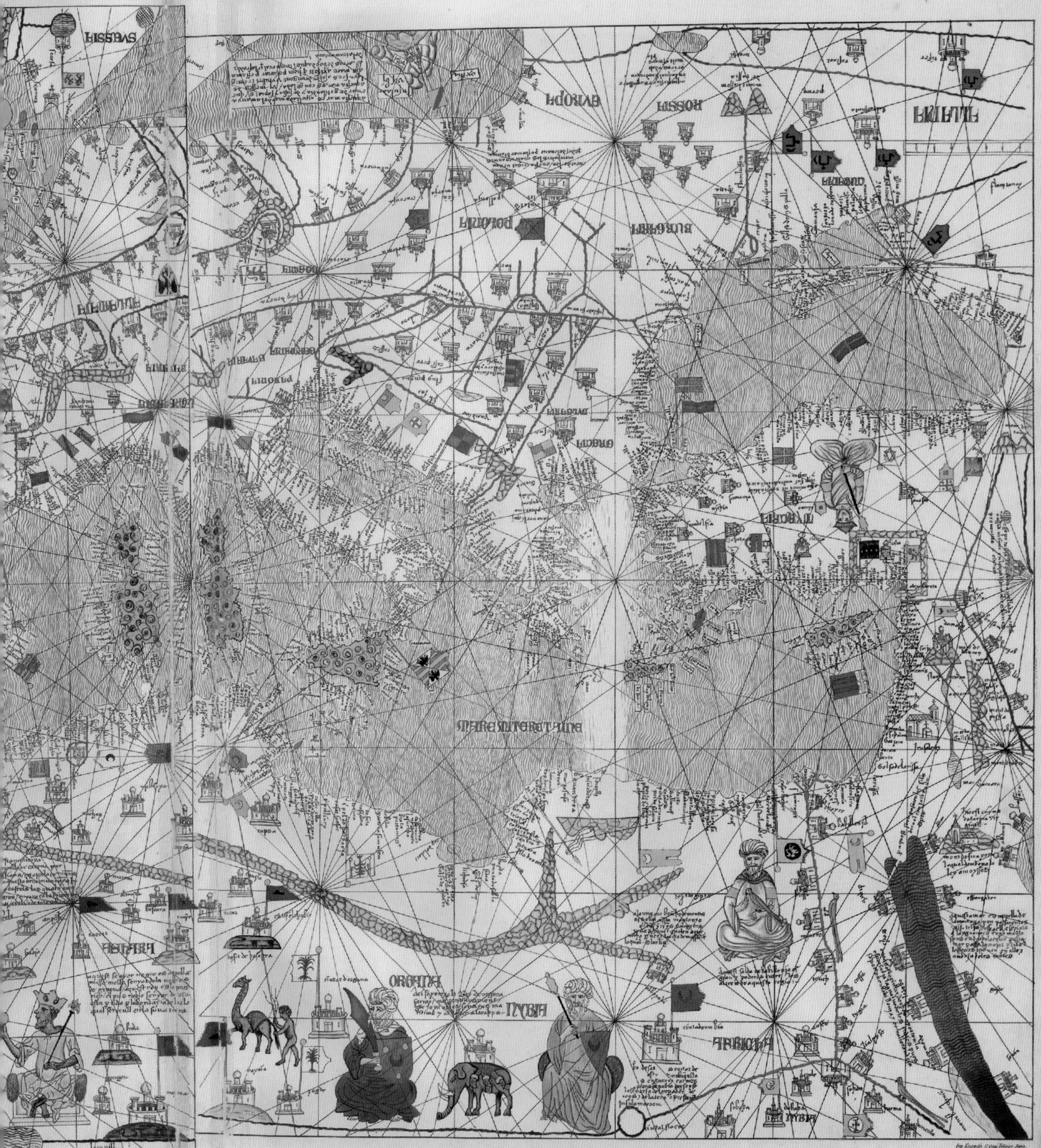
EVROPA
ROSSIA
POLONIA
BVRGARIA
ORGANA
NVBIA

*Imp. Kaeppelin 17, Quai Voltaire, Paris.*

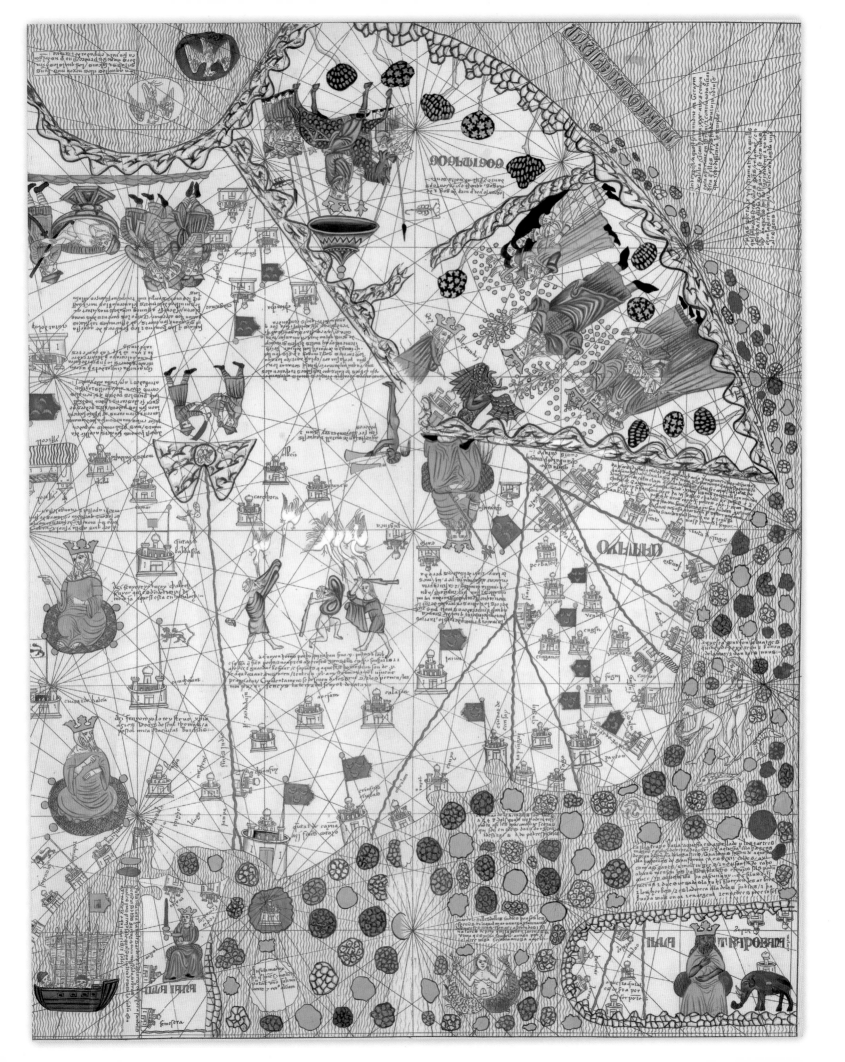

# CHAPTER 2

# Cartography in the *Age of Discovery*
## (c.1450–c.1650)

Right: Abraham Ortelius's world map from *Theatrum Orbis Terrarum*, the first modern geographical atlas. The edition of the atlas from which this map comes was published in Antwerp in 1603, five years after the death of Ortelius. The thirty-first and final edition was published in 1612, by which time the number of maps had increased from the original 70 in the first atlas of 1570 to 167.

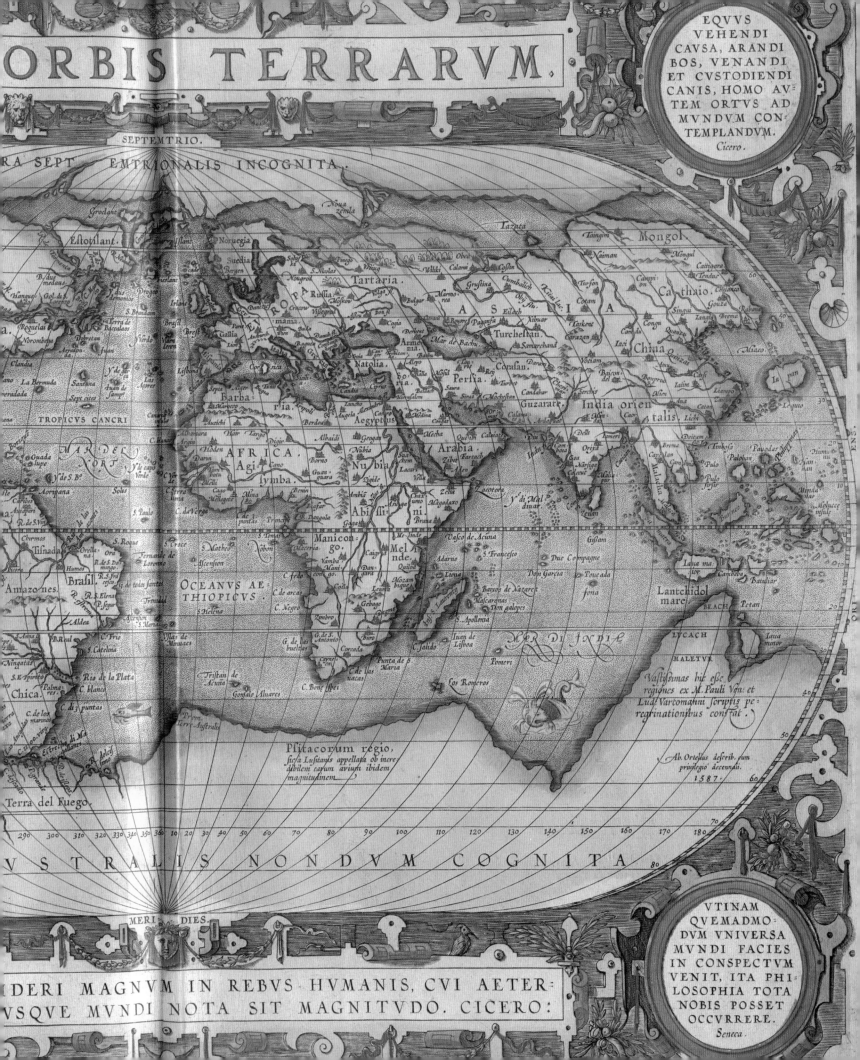

# ORBIS TERRARVM.

EQVVS
VEHENDI
CAVSA, ARANDI
BOS, VENANDI
ET CVSTODIENDI
CANIS, HOMO AV-
TEM ORTVS AD
MVNDVM CON-
TEMPLANDVM.
Cicero.

SEPTEMTRIO.

...RA SEPT EMTRIONALIS INCOGNITA.

Groclant

Estotilant.

Noua zemla

Tazata

Taingim   Mongol

Naiman   Mongul   Catrigara

Noruegia   Tendur   60

Suedia   Turfon   Cathaio   Chiorza

Bergen   Tartaria.   Grustina   Cambalich   Cotam   Gouza   Rabana

EVROPA   Russia   ASIA   Turfon   Camul   Congu   Singui   Tangut   Brema

Irlac   Moskov   Obij flu   Taskent   Iaci   China   Quizat   Miaco

Gallia   Armenia.   Samarchand   Corazan   Yociam   Aua   Cheli   Liampo   Lequio

Lisbona   Natolia.   Persia.   Darum   Chari   Baicou   Iatim   30

Corica   Candia   Cypro   Saura   Turfet   Candabar   Serchis   India orien   talis   Lichi

TROPICVS CANCRI   Barba   ria.   Aegyptus   Arabia   Deli   Pameri   Dieicam

MAR DEL   Hoden   Digir   Albaidi   Geogan   Mecha   Quilim Caliata   Orixa   Timhola   Pauodas   Philippinas   20

NORT   Darun   Borno   Nubia   Suachem   Zibet Fartach   Goa   Calicut   Malua   Pulo   Palohan

AFRICA.   Cano   Guan   guara   Nubia   Lacuri   Narsiga   Maldiua   Pulo   furfur

Agi   lymba.   Benin   Zeila   Zacotora   Zeilan   Molucce insule

S. Paulo   C. da Verga   Abiss   ni.   Chax   umo   Magadaxo   Y di Mal   diuar   Mindu   dao

Chirmos   Manicon   Me linde   Vasco de Acuna   Gissam   Iaua ma   ior   Cambaba

Tisnada   go.   Melinde   Adarno   S. Francesco   Due Compagne   Batudiar

Brasil   Vamba   Caigo   Quilod   Liona   Don Garcia   Toueada   Lantchidol   Petan   mare

OCEANVS AE   Mani   Dangara   Mozam   bique   Baixos de Nazaret   fona   BEACH   20

THIOPICVS.   S. Helena   Gebage   Mascaranas   Don galopes

Amazones   G. de las   bueltas   S. Apollonia   MOR DI INDIA   LYCACH   Iaua   minor

Corcada   Iuan de   Lisboa   MALETVR   30

Chica   Punta de   Maria   Pomeri   Los Romeros

Vastissimas hic esse
regiones ex M. Pauli Ven: et
Ludi Vartomanni scriptis pe-
regrinationibus constat.

Tristan de
Acuna   Gonsalo Aluares

C. de las   uacas   C. Bone Spei

Ab. Ortelius describ. cum
priuilegio decennali.
1587.   60

Terra del Fuego.

Psitacorum regio,
sic/a Lusitanis appellata ob incre-
dibilem earum auium ibidem
magnitudinem.   50

290   300   310   320   330   340   350   360   10   20   30   40   50   60   70   80   90   100   110   120   130   140   150   160   170   180   60   70

...VSTRALIS NONDVM COGNITA   80

MERIDIES.

...DERI MAGNVM IN REBVS HVMANIS, CVI AETER-
...VSQVE MVNDI NOTA SIT MAGNITVDO. CICERO:

VTINAM
QVEMADMO-
DVM VNIVERSA
MVNDI FACIES
IN CONSPECTVM
VENIT, ITA PHI-
LOSOPHIA TOTA
NOBIS POSSET
OCCVRRERE.
Seneca.

# New Ideas, New Technology and New Men

Left: A screw press of the type used by Johannes Gutenberg in the mid-fifteenth century, when his new concepts revolutionized the world of printing.

THE EARLY FIFTEENTH CENTURY WAS ONE OF THOSE RARE PERIODS in history when the combination of knowledge, inspiration and the right individuals hastened progress at such a rate as to herald the dawn of a golden age, which in this case was in exploration and cartography.

The century began with western Europe's rediscovery of Ptolemy's *Geographia*, which was brought from Constantinople and translated into Latin in about 1405, and thereafter into French and Italian. European map-makers were now able to learn the fundamentals of Ptolemy's mathematical cartography, including his methods to calculate geographical positions, establish coordinates and use map projections. Ptolemy's work also influenced the regular use of a north orientation.

Shortly thereafter, developments in printing transformed the way that information was reproduced and disseminated. Paper had been invented in China around AD100, and its manufacturing process had been adopted by the Arabs in the eighth century and in Islamic Spain by the tenth. By 1400, paper was being manufactured in Italy, France and Germany.

Around 1440, the German goldsmith Johannes Gutenberg adapted a number of existing technologies in order to develop a revolutionary system for printing. He used a press based on earlier screw presses, devised a hand-mould that allowed the setting up of movable type and created his pieces of type out of a metal alloy. He also overcame the challenges of traditional water-based inks – which blotted and spread too widely on paper – by developing an oil-based ink. Many early printed maps were produced from woodcuts, and after Gutenberg's

contributions metal lettering into the woodcuts to produce were a combination of the best technologies. The English philosopher Francis Bacon later wrote that these advances had "changed the whole face and state of things throughout the world".

Meanwhile, navigational techniques had greatly improved thanks to the development of new instruments. Most importantly, the simple compass, which consisted of a magnetized pointer in a bowl of water and had been used in Europe since the twelfth century, was replaced by the dry mariner's compass. This compass comprised a freely pivoting magnetized needle suspended over a card marked with compass points, all held in a covered box that could be fastened in line with the keel of the ship. As the ship changed direction, the card would turn and therefore the course of the ship would always be indicated.

The second major advance in instruments was the mariner's astrolabe (distinct from the astrolabe proper), which was designed for use on ships in heavy seas or strong winds. A brass, ring-like implement, the mariner's astrolabe allowed the user to determine latitude at sea by measuring the Sun's noon declination or the meridian altitude of specific stars.

During the fourteenth century, interest in travel had increased greatly throughout Europe with the publication of (initially in Old French) *Livres des merveilles du monde*, now commonly called *The Travels of Marco Polo*, which related the Venetian's epic journey

Above: A cleverly designed fifteenth-century box containing both a sundial and a compass. The early compass card is not nearly as elaborate as later ones would be.

Opposite: The *mappa mundi* of the Camaldolese monk Fra Mauro, which was produced at the monastery of San Michele on the island of Murano in the Venetian lagoon.

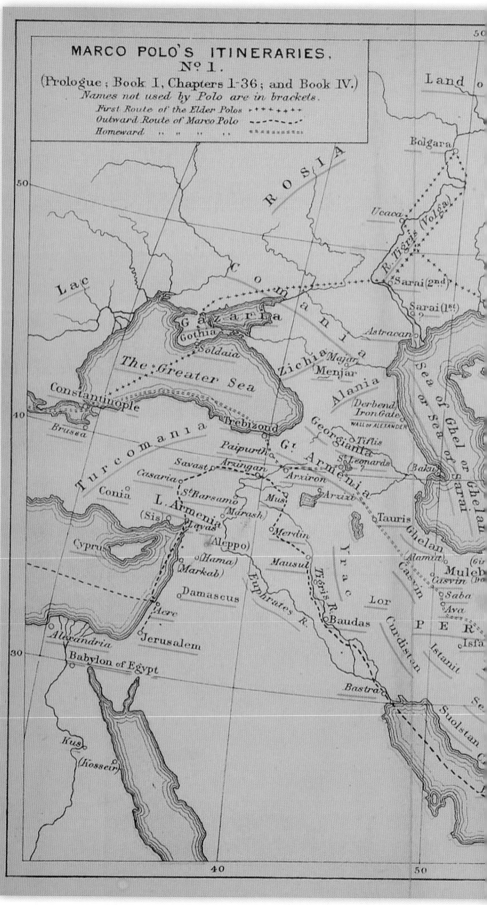

to Central Asia and China. But the man whose influence was the key component in beginning the Age of Discovery was a prince of Portugal who became famous as "Henry the Navigator" (1394–1460). In his twenties, Henry began to dispatch expeditions to explore the nearby stretches of the Atlantic Ocean. Later, he sent out vessels to reach India by following the unknown coastline of West Africa. Henry hoped to foster commerce, especially that of the lucrative spice trade. An important element of his success was the development of caravels – light, fast and highly manoeuvrable sailing ships, which were rigged with lateen (triangular) sails that gave them the capacity to beat against the wind (sail to windward) and therefore made them better able to reach previously inaccessible areas.

An early result of Henry's programme was the rediscovery of the Atlantic Ocean archipelago of Madeira in 1419 (which had long before been visited by the Genoese). As the islands were settled, sugarcane was introduced and the classic plantation regime for the sugar industry was first installed. Henry's captains also made their way down the coast of Africa and by 1446 they had reached the Gambia river. By 1456 the Cape Verde Islands had been discovered, and during Henry's lifetime Portuguese sailors had progressed as far as present-day Sierra Leone. These contacts initiated the infamous trade in gold and slaves.

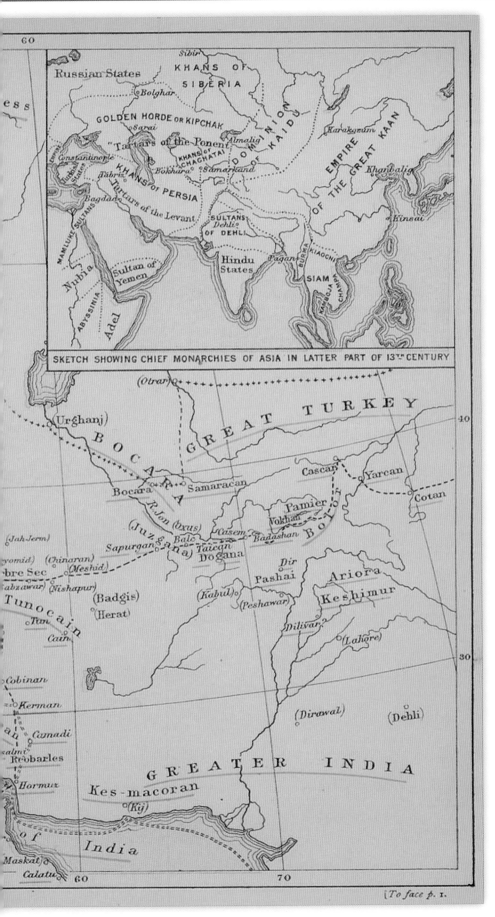

Opposite above: Henry the Navigator, as portrayed in the third panel of the altarpiece painted for the convent of São Vicente by the fifteenth-century Portuguese artist Nuno Gonçalves.

Left: A map of Marco Polo's early journeys, from an eighteenth-century British edition of the book *The Travels of Marco Polo*. Fra Mauro included all of Polo's place-names in his own map of the world.

In the final years of Henry the Navigator's life, Fra Mauro, a Venetian monk commissioned by Henry's nephew King Afonso V (reigned 1438–81), produced what is considered to be the last of the great *mappae mundi*. A circular giant with a diameter of nearly 2 m (6 ft 4 in), Mauro's map not only included information about Asia (gathered from Marco Polo's account) but also outlined the known coasts of both East and West Africa. Although the southern parts of Africa were not accurately represented, this famous map encouraged the belief that the riches of the east could be reached by sailing around the continent.

The Portuguese continued their exploration after Henry's death. In 1471 the island of São Tomé was discovered in the Bight of Biafra. São Tomé had the perfect climate and soil for sugar, as well as easy access to slave labour, and it was there that the model for the slave plantations of the New World was perfected. With the building on the Gold Coast of São Jorge da Mina (Elmina, present-day Ghana) in 1482, the Portuguese established the first European fortress in tropical Africa, which became a major trading post. The huge financial success of the trade in gold and slaves encouraged the Portuguese to continue down the coast of Africa in search of further lucrative trading centres and operations.

# Columbus and the New World

Above: The Erdapfel, the oldest extant globe. As it was produced before Columbus returned from the New World, the Americas are not shown, and a vast ocean extends between Europe and Asia.

Above centre: A portrait of Vasco da Gama. The explorer twice returned to India after his first great voyage, dying of malaria in the city of Cochin (now Kochi) on Christmas Eve 1524.

Below: A statue at the Bartolomeu Dias museum complex in Mossel Bay, South Africa. The statue of the Portuguese explorer commemorates his landing and first contact with the indigenous people.

**A**S THE END OF THE FIFTEENTH CENTURY approached, the Portuguese were virtually unchallenged in maritime exploration. In 1455, a papal bull (*Romanus Pontifex*) by Pope Nicholas V (papacy 1447–55) confirmed Portugal's dominion over all the regions its navigators had discovered. In addition, other Christian nations were forbidden from infringing the Portuguese crown's rights of trade and colonization in those lands, which meant that the investigation of the African coast by the Portuguese therefore continued unfettered. In 1482, Diego Cão discovered the River Congo, and four years later he reached what is now known as Namibia.

In 1488, the Portuguese nobleman Bartolomeu Dias reached the Cape of Good Hope, rounded the tip of Africa and established that the coast turned northeast – a maritime route to India, the Indian Ocean and beyond had been opened. A decade later, Vasco da Gama passed around southern Africa, sailed up the east coast to present-day Kenya and turned east to reach India. Da Gama then returned to Portugal with a cargo worth many times the cost of the expedition, having disproved conclusively many ancient theories regarding the shape of the Earth's landmasses.

While the Portuguese relentlessly pursued their round-Africa course to Asia, others considered alternative routes to the silk, spices and other valuable goods from the East. With the fall of Constantinople to the Ottoman Turks in 1453, Marco Polo's land route, as well as the trek from the Mediterranean to the Red Sea, became dangerous in addition to difficult. Several decades later, Christopher Columbus, a Genoese navigator, devised a totally different plan, which he spent years attempting to convince King John II of Portugal (reigned 1481–95), and subsequently the Spanish "Catholic Monarchs" King Ferdinand and Queen Isabella I, to sponsor.

Columbus believed that India and other Asian lands could be reached by sailing west across the Atlantic in a voyage that would be shorter than one entailing a journey around Africa. His theory was correct – that Asia could eventually be reached by sailing westward – but the distances involved were massively underestimated. Columbus's calculation of the circumference of the Earth was incorrect, as was his assumption about the size of the Eurasian landmass and where Japan lay in relation to the Asian mainland. The latter two errors may have been due to the influence of several maps that had recently been published. Around 1474, Paolo Toscanelli, a Florentine mathematician and astronomer, had produced a map to accompany his own plan to sail westward to the Indies. Toscanelli later gave a copy to Columbus, who took it with him on his first voyage to the New World (*novi orbis*, as it became known), having never realized Toscanelli's miscalculations.

While Columbus was seeking a patron, Henricus Martellus, a German cartographer working in Florence, produced a world map that was heavily influenced by Ptolemy. However, this map also severely underestimated the distance involved in a westward sea voyage to Asia, and it seems likely that it was incorporated into Columbus's plans because it bears significant similarities to a map drawn in 1490 in the Lisbon workshop of Columbus and his brother Bartolomeo. It was key to Columbus's argument that Martellus's map showed the tip of Africa to be much farther south than it actually was, because it suggested that such a journey would be longer than his proposed voyage to the west.

Another important record of the cartographic thinking of the time is a globe that was produced in around 1492 under the direction of the German cosmographer Martin Behaim. The oldest extant European globe, the *Erdapfel* ("Earth Apple") as Behaim called it – it is also known as the Nuremberg globe – was constructed of a metal

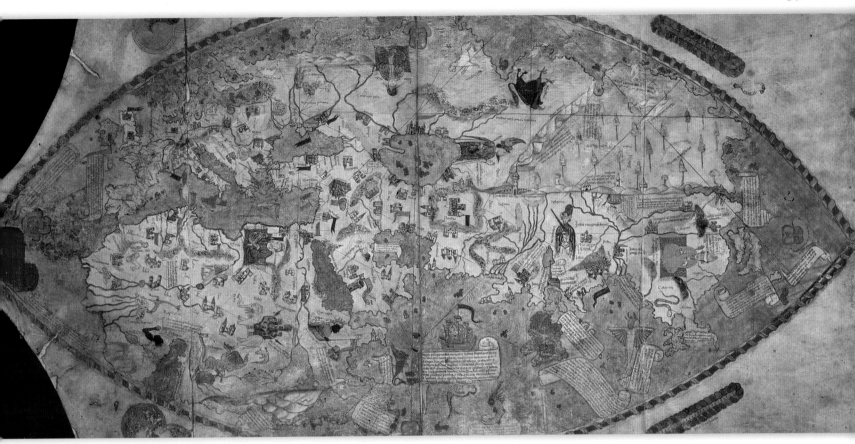

Above: Produced around 1457 with an unusual elliptical shape, this map is frequently known as "the Genoese Map". In the 1940s it was suggested that this was the map that Paolo Toscanelli sent to Columbus, emphasizing the possibility of a western sea route to Asia, but there is no strong evidence for the claim.

ball overlaid with a map painted by Georg Glockendon. A masterpiece of production, the globe represents an intriguing combination of Ptolemy's theories and the latest information from Portuguese voyages. It bore approximately 1,100 place names, information about European monarchs and improvements of lines of latitude and longitude. But it also located Japan only 2,400 km (1,500 miles) from Europe, corresponding in that way with Columbus's geographical distances.

Despite his mistakes, Columbus was to lead four voyages to the region of the Caribbean Sea, and his efforts changed the maps of the world forever.

An INDIAN CACIQUE of the ISLAND of CUBA, addressing COLUMBUS concerning a future state.

## Ptolemy's Error and Columbus

Ptolemy influenced not only Roman cartography, but also map-making in the Islamic world and medieval Christian Europe. However, Ptolemy did make fundamental mistakes and these proved to have far-reaching effects. He underestimated the size of the Earth and indicated that Europe and Asia combined to cover half the planet's circumference, whereas in actuality these two continents extend to only 130° (rather than 180°). So eminent was Ptolemy that, 13 centuries later, Columbus used this information to plan his own western voyage, which resulted in proving the fallibility of Ptolemy's estimates.

But it was not only Ptolemy who led Columbus astray. The navigator believed the estimates produced in the ninth century by the Persian astronomer Alfraganus (Ahmad ibn Muhammad ibn Kasir al-Farghani), which indicated that a degree of latitude at the equator was equal to approximately 91 km (56.7 miles). However, Columbus did not realize that Alfraganus was working in Arabian miles (about 1,830 m) rather than Italian miles (about 1,240 m). This caused Columbus to underestimate the circumference of the Earth and overestimate how far east Asia extended. It is ironic that Columbus owed his success to these two significant errors.

Left: A native leader and his people approach Columbus on the shores of Cuba. The explorer's first landing on the island was on 28 October 1492.

# The Spanish and Portuguese:
## Mapping Recent Discoveries

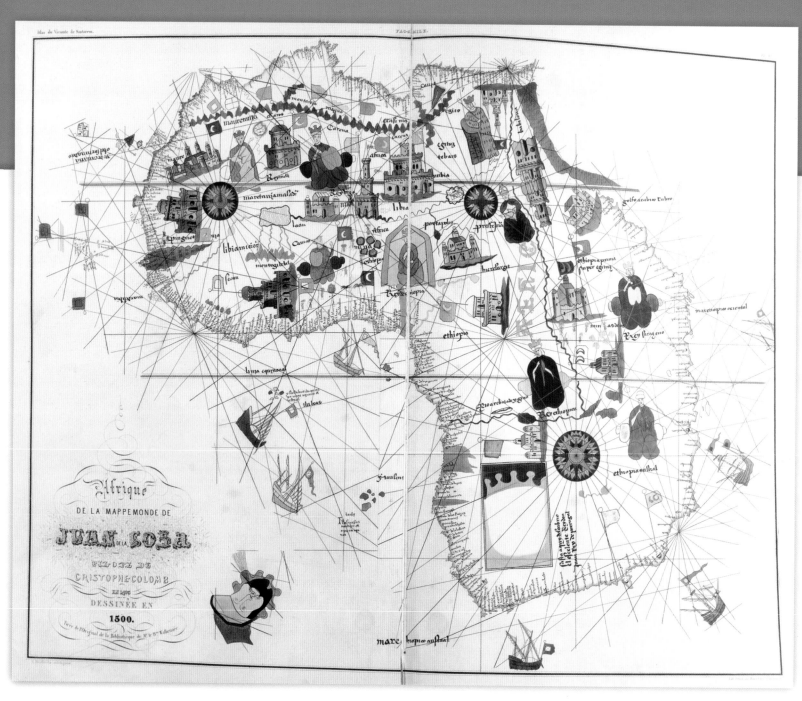

COLUMBUS'S FIRST VOYAGE IN 1492 began one of the greatest rivalries in the history of exploration and conquest. After his return to Spain in 1493, a papal bull guaranteed Spain's rights to all the newly found lands, and the next year the Treaty of Tordesillas divided any forthcoming discoveries between Spain and Portugal along a line of meridian 370 Castilian leagues – approximately 2,060 km (1,280 miles) – west of the Cape Verde Islands. All territories to the east of the demarcation would belong to Portugal; those to the west would go to Spain. Within a decade, Spanish and Portuguese ships in abundance were sailing west

Above: A nineteenth-century copy of the African segment of Juan de la Cosa's parchment map of the world, which he created in 1500. Like the original, it includes illustrations of important rulers and buildings.

to enable discovery, conquest, the spread of Roman Catholicism and the acquisition of riches and fame by those brave and audacious enough to seize the moment.

This great wave of exploration had a huge impact on cartography, and a vast number of maps began to be produced, recording both the newly discovered regions and the world as a whole. One of the first was compiled by Juan de la Cosa, who had accompanied Columbus on three of his voyages. In 1499 he served as the pilot for Alonso de Ojeda during a journey in which they reached the coast of the South American mainland at what was to become Guyana and then followed it northwest to present-day Colombia. After returning to Spain, de la Cosa produced the first-known map to show both Europe and the New World. Of particular interest are his outline of Cuba, which Columbus did not believe to be an island, and his recording of Pedro Cabral's landfall in what is now called Brazil, John Cabot's voyage to Newfoundland, or Terra Nova, and Vasco da Gama's route to India.

## The Aztec Mapping Tradition

In 1519 Hernán Cortés, a conquistador who had fought in the conquest of Isla Juana (now named Cuba), landed on the coast of what is now Mexico with 500 men, 16 horses and six cannon. He made his way inland towards the great city of Tenochtitlan – located on an island in a lake – from where Montezuma II (reigned 1505–20) ruled over the Aztec Empire. The Spaniards seized Montezuma, who became a Spanish puppet, but five months later, when Cortés left the city, the Aztecs rebelled and proceeded to kill him. Cortés thereupon sacked Tenochtitlan and established a ruthless Spanish rule.

Even before reaching the capital, Cortés realized that the native culture possessed remarkable map-making skills. He was presented with indigenous itineraries produced on henequen cloth (made from the fibre of the henequen plant), on which were marked pueblos, estates and roads. The Spanish also found numerous other maps, including those of local communities that were accompanied by narrative histories, drawings of trade routes, property charts and celestial maps that were far more advanced than those produced in Europe.

The conquering Spanish worked with Aztec map-makers to record many elements of their new empire. One example of this cooperation was the map of Tenochtitlan as it was before the Spaniards destroyed the city in 1521. Cortés sent this, in 1524, to King Charles I of Spain (reigned 1516–56, and as Holy Roman Emperor Charles V 1519–58). Although it is not known who drew the map, it was either the work of Aztec cartographers or based on their originals. Both Aztec and Spanish influences can be seen in the map, to which Latin text was added before it was prepared in a woodblock format in Nuremberg.

Left: A map of Tenochtitlan, the Aztec capital built on an island in Lake Texcoco, showing an idyllic image of the region before the Spanish conquest.

Ojeda and de la Cosa were accompanied on their journey by the Florentine Amerigo Vespucci, who parted from them in Guyana and sailed south, down the coast of Brazil. The claim that Vespucci participated in several other expeditions is contentious, but it is beyond dispute that he took part in a voyage in 1501–02 that convinced him that the recently discovered lands were not part of Asia but a "new world".

After his return to Europe, Vespucci worked in Seville for La Casa de Contratación (The House of Trade) – the Spanish government's overseer of commerce with the new colonies – in a department responsible for producing the Padrón Real (Royal Register), Spain's official, secret master map of the previously unknown territories.

The Portuguese crown established similar departments, including a covert one to chart all new geographical discoveries, which were closely guarded in order to maintain the country's monopolies. However, in 1502, Alberto Cantino, an agent for the Duke of Ferrara, had Portugal's primary chart – the Carta Padrão del el-Rei – copied and smuggled to Italy. There, to the outrage of the Portuguese, a map that became known as the Cantino Planisphere was produced on three large sheets of parchment. It was an extraordinary document of current world knowledge, which charted with remarkable accuracy the coasts of Africa and India; it also depicted, for perhaps the first time ever, the coast of Brazil and placed openly on a map the line agreed in the Treaty of Tordesillas.

Some of the information gleaned from the Cantino Planisphere was reproduced in subsequent years, which suggests a number of cartographers were privy to it. Three significant maps using the planisphere's data quickly appeared: in Italy, the Low Countries and Germany. In 1506, the Italian cartographer Giovanni Contarini and the engraver Francesco Rosselli put out the first printed world map to show the New World. But the Contarini map was not widely circulated, and therefore not nearly as well known as one that was produced in 1507 by Johannes Ruysch of Utrecht.

One of six supplemental maps in a new edition of Ptolemy's *Geographia*, Ruysch's fan-shaped effort is notable for both its accuracies and inaccuracies. Among the former were the detailed outlines of the coasts of Africa; the inclusion for the first time of the islands now known as Sri Lanka, Madagascar and Sumatra; and an indication that some of the western discoveries comprised a "new world". Less accurate was Ruysch's depiction of Greenland and Newfoundland as parts of Asia (previously both islands had been shown as part of Europe), the inclusion of four fictitious islands around the North Pole and his indication that "Spagnola" (Hispaniola – present-day Haiti and the Dominican Republic) was the same as Marco Polo's "Sipganus" (Japan).

In 1507, the same year that the Ruysch map appeared, Martin Waldseemüller, a German cartographer working in Saint-Dié in what is now France, created arguably the most influential map in the history of cartography as far as place names are concerned. Produced to accompany a treatise entitled *Cosmographiae Introductio* (*Introduction to Cosmography*), Waldseemüller's map of the world comprised 12 contiguous woodblock prints that measured nearly 1.4 m × 2.5 m (4 ft 6 in × 8 ft 2 in).

The map has numerous remarkable features. North of the Caribbean island groups discovered by Columbus is a peninsula not dissimilar to Florida – yet Waldseemüller's map predates Juan Ponce de León's discovery of Florida by six years. More importantly, this was the first map to show the recent finds as individual continents separated from Asia by a wide ocean – and this was half a dozen years before Vasco Nuñez de Balboa became the first European to see the eastern reaches of the Pacific Ocean and 15 years before the survivors of Ferdinand Magellan's circumnavigation returned to Spain, after having become the first men to cross the Pacific.

The reason why Waldseemüller's map showed the two new continents, one north of the other, has long been debated. It has been argued that he realized the differences in the longitudes proposed by Ptolemy and those reported by recent explorers meant that there must have been more uncharted territory between Asia and the new lands; alternatively, it was simply an accurate leap of intuition on his part. Waldseemüller was certainly familiar with two letters that have (controversially) been attributed to Vespucci, which were published and widely circulated in the first years of the sixteenth century. These notes promoted the idea that the South American coast was actually part of a new continent.

This concept won over Waldseemüller, and the new southern continent on his map was awarded the name "America". Explaining his use of the feminine, Latinized version of Vespucci's first name, Waldseemüller wrote: "I do not see why anyone would rightly forbid calling it after Americus who discovered it ... Amerige, that is, land of Americus, or America, since both Europa and Asia received their names

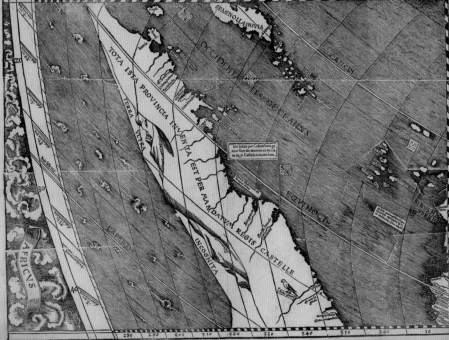

from women." Although Waldseemüller removed the name from his 1513 atlas, it had by that time been noted for inclusion on other maps – such as those produced by Johannes Schöner in 1515 and Petrus Apianus in 1524 – and had become accepted.

A reported 1,000 copies of Waldseemüller's map were printed, but within a few centuries every copy had disappeared. Then, in 1901 while conducting research at Wolfegg Castle in Baden-Württemberg, the Jesuit scholar Josef Fischer found a portfolio of materials that had been used by Schöner, and it contained a copy of Waldseemüller's masterpiece. This is the only known copy in the world, and in 2003 it was purchased by the Library of Congress in the United States.

Throughout the sixteenth century, Spanish and Portuguese penetration into the New World – and, in the case of Magellan, around the Earth – was reflected in progressively detailed maps. Diego Ribero of La Casa de Contratación produced several notable maps, including, in 1529, the first version of the Padrón Real to include the details gathered on Magellan's circumnavigation (1519–22), which therefore made it the first map to show the truly vast extent of the Pacific Ocean.

For almost 30 years, Battista Agnese, a Genoese cartographer working in Venice, produced a series of atlases, of which more than 70 still survive. The atlases tended to include one world map and an additional 12 to 14 regional ones in portolan style. Typical of this format was his 1544 atlas dedicated to the abbot of St Vaast, in which was marked Magellan's complete navigational route. Agnese also inscribed in gold the passage from Cadiz to Peru that the Spanish treasure ships followed, including the overland portage across the Isthmus of Panama. Agnese's maps were also the first to show Baja California as a peninsula separated from Mexico by the Gulf of California. Eight decades later, in a retrograde step, this area – and, indeed, all of California – would be regularly mapped as an island by the great Dutch cartographers.

Opposite: An illustration of Amerigo Vespucci, shown holding a map of the Americas. We know that the portrait was produced after his death in 1512 because the map gives details that were only discovered in later years.

Left: Part of the three left-hand panels from Martin Waldseemüller's world map of 1507. On the bottom section appears his introduction of the term "America", first used for the southern of the two continents of the New World.

Below: A 1529 version of Diego Ribero's Padrón Real, showing the discoveries known at the time by Spain's La Casa de Contratación. Like the great voyager Ferdinand Magellan, Ribero was originally Portuguese, but served Charles I of Spain.

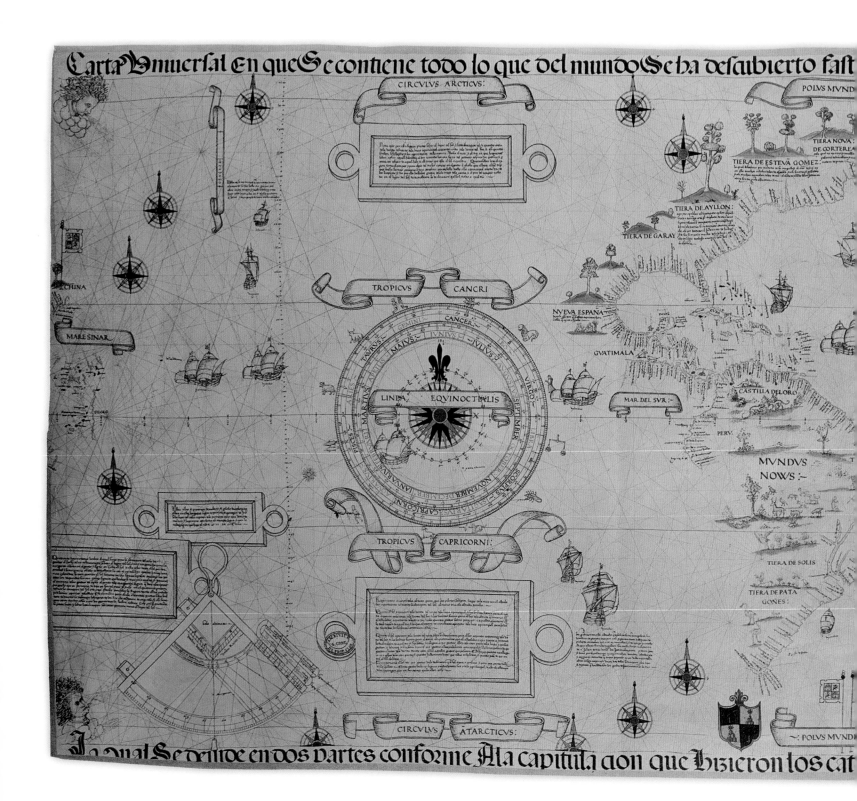

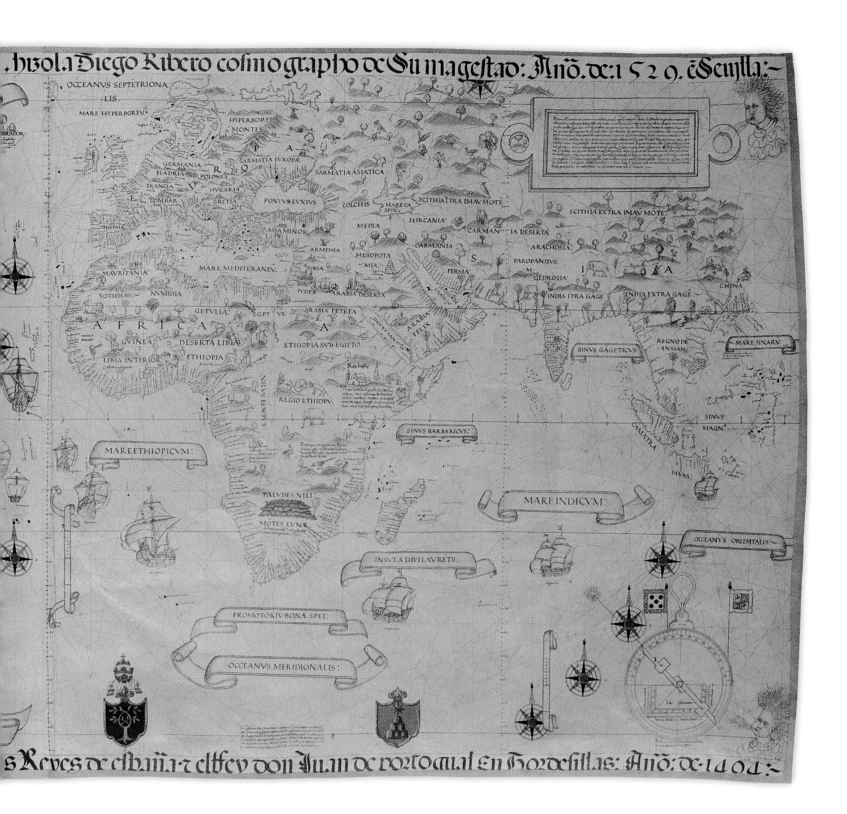

# Cosmographies
## and the Development of
# Projections

**T**HROUGHOUT THE AGE OF DISCOVERY (*c.*1450–*c.*1650) new geographical information was constantly being added to maps, and both the appearance of maps as well as the uses to which they were put were rethought and changed.

In the early sixteenth century there was a growth of interest in cosmography, which was then defined as the science dealing with the structure of the universe and its constituent parts – geography, geology, astronomy, history and natural history. The textbooks dealing with various aspects of those subjects, and illustrated with maps

and figures, were known as *cosmographiae* and became increasingly popular. One of the first, published in 1524, was *Cosmographicus Liber* (*Book of Cosmography*) by Petrus Apianus, which was eventually translated into 14 languages. An introduction to scientific knowledge based largely on Ptolemy, the book had little original information but it was lavishly illustrated.

Within two decades, Sebastian Münster of the University of Basel had produced two major works: *Geographia Universalis* (1540), which is often considered the high point of Ptolemaic revisions, and *Cosmographia Universalis* (1544), one of the first works of geography designed for public consumption. By 1628 the latter had gone through 36 editions, demonstrating the widespread eagerness for scientific and geographical knowledge. Münster was the first cartographer to display the four known continents (Europe, Asia, Africa and the New World) as individual maps.

The physical appearance of world maps also changed with the development of new cartographic projections designed to allow the transfer of three-dimensional information about the surface of the Earth (or a celestial sphere) onto a flat surface. Projections are mathematically definable rules for establishing points in relation to a "graticule" – the grid of lines representing east-to-west parallels of latitude and north-to-south meridians of longitude. Ptolemy had been familiar with the distortions of shape, area, distance, direction or scale that can arise in producing a flat map, and had attempted to resolve the problems by using conic and conic-like projections, in which the parallels are arcs of circles and the meridians are straight lines converging at a point after the distance between them is consistently reduced by a fixed factor. For more than a millennium thereafter, projection proved of little interest to most cartographers, because maps in both Europe and the Islamic world reflected philosophical or religious agendas as much as the accuracy of geographers.

During the Age of Discovery the standard projections became more problematic because of the increasing size of the known world, and the period was marked by new projections as map-makers reflected in their work the fact that cartographical preciseness had become more important to rulers, administrators, military leaders, merchants and explorers. At least 23 map projections were used during this period, including 14 that were invented after 1426. One of the first was the Donis projection, a trapezoidal one used by Donis Nicolaus Germanus in a world map for a 1466 version of Ptolemy's *Geographia*.

Opposite above: A woodcut of a world map from Petrus Apianus's *Cosmographicus Liber* of 1524. The map's use of the term "America" helped make certain that that the name continued in vogue.

Opposite below: A map of Africa, originally prepared by Sebastian Münster for his *Cosmographia Universalis*. This edition was published in 1572, 20 years after Münster's death.

Below: Johannes Ruysch's map of the world, produced for a 1507 edition of Ptolemy's *Geographia*, published in Rome. The map is an example of Ptolemy's classic conic projection.

Overleaf: A hand-coloured engraving of a map of Europe produced by Gerardus Mercator in approximately 1554. This was produced shortly after he moved to the city of Duisburg in the Duchy of Cleves.

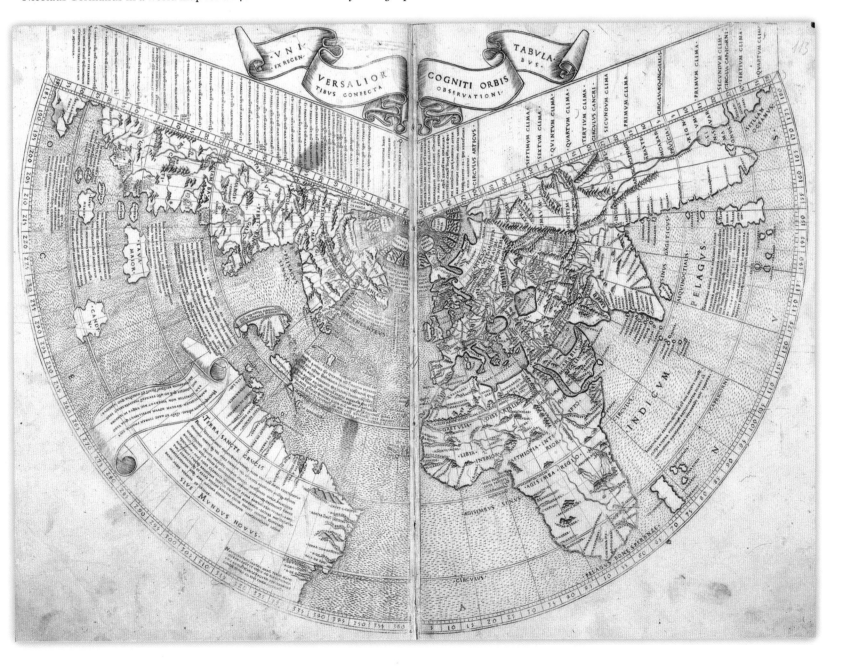

EVROPA,
ad magnæ Europæ Ge:
rardi Mercatoris P. imitati:
onem, Rumoldi Mercatoris F.
cura edita, feruato tamen
initio longitudinis ex ratio:
ne magnetis, quod Pater
in magna sua vniuer:
sali posuit.

Medius Meridianus 50. reliqui ad hunc
inclinantur pro ratione 60. & 40.
parallelorum.

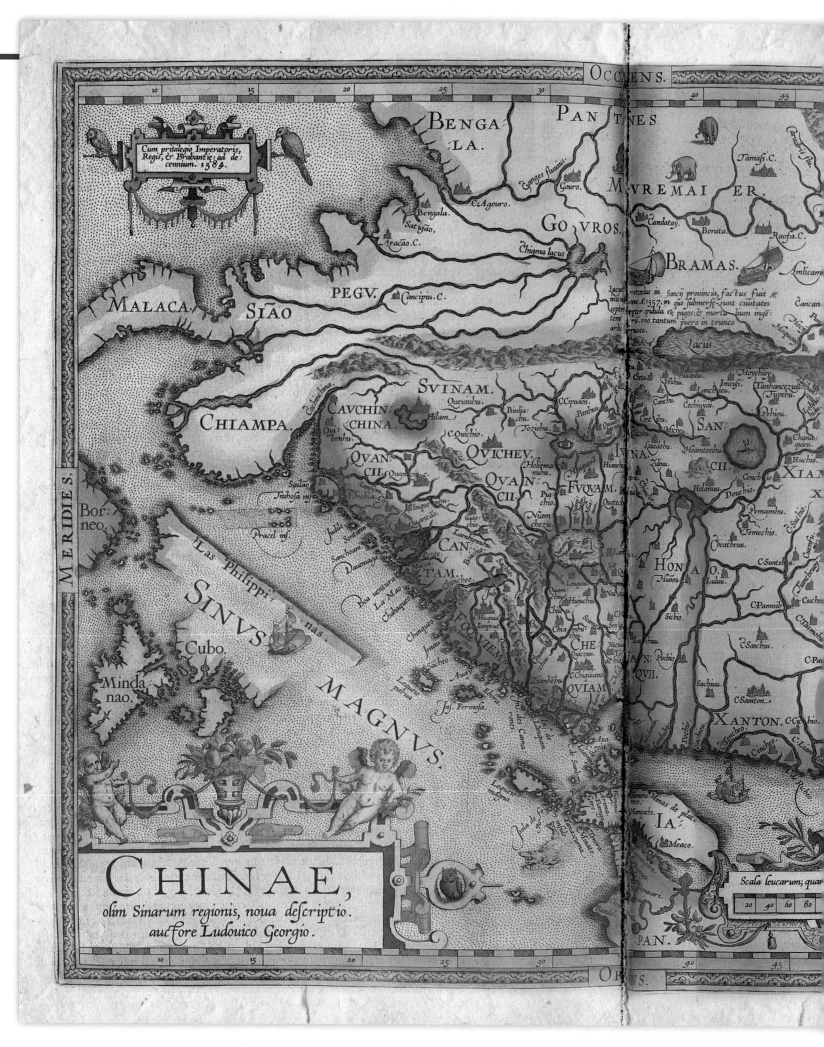

10     15     20     25     30     40     45

PAN TNES

BENGA-LA.

*Cum priuilegio Imperatoris,*
*Regis, & Brabantiæ ad de-*
*cennium. 1584.*

Ganges fluuius.

Gouro.

MOVREMAI ER.

Tamafi.C.

C. Agouro.

Bengala.

Satigao.

Aracão.C.

Candatay.

Borata.

Raofa.C.

GO VROS.

Chiama lacus.

BRAMAS.

Amlicam

PEGV. Cancipiu.C.

*Lacus rotundus in Sancij prouincia, factus fuit*
*mutatione A.1557. in quo submersæ sunt ciuitates*
*septem præter oppidula et pagos: et mortuum ingē-*
*tem numeru, uno tantum puero in trunco*
*arboris seruato.*

Cancan

Hochi

Haquay

MALACA. SIÃO

Lacus.

SVINAM.

Queuauhu.

Hoychiou.

Veuhu.

Putandu.

Imtuſi.

Lanchiou.

Tanhancezua.

Jupeiu.

Pehiou.

CHIAMPA.

CAVCHIN-CHINA.

Hilam.

C.Quichio.

C.Cipuaon.

Panhau.

Nachan.

Cancha.

Micheu.

Cochuyai.

Chinchina.

Quirbenhu.

QVICHEV.

SAN.

Chana-quieu.

QVAN CII. Quanciu.

Holiema nuhu.

Hianchu.

Hfiungxia.

Laadohu.

Hoamtenhu.

Huchu.

QVAN CII.

Pia-chio.

FVQVAM.

Quoteh.

Zidiui.

CII.

XIAN

Cench io.

Holanu.

Douchio.

Bor-neo.

Sailao.

Inthofa inf.

Sinchina.

Niam chezu.

Yemamhu.

Temechio.

C.Suchio.

Cuenfi

Pracel inf.

Judila.

Santami.

Inquo.

Lui fohhi.

HON A O.

Huiou.

Lulu.

C.Suntehu.

Tanchengy

Cuche

*Las* Philippi nas

Sanchoam.

Duenuga.

Le Mao.

Chabaquco.

Chia

Quam chiu.

Huuchu.

Sen

Nod fan.

Sichio.

C.Pamnih.

C.Pa

SINVS

Cubo.

Boa uentura.

Chinquco.

Imai.

TAM.

Macao.

Chincheu.

Huqua

Luupe ui

Qua

CHE

Quacou.

NAN Pochio.

C.Saichiu.

C.Pa

Minda nao.

MAGNVS.

Sacheo.

Litera paſſua

Inf. Fermoſa.

Aua

Ziankehu.

C.Chiquano.

QVIAM.

QVII.

Sachuu.

C.Samton.

XANTON.C

Cochio.

Pochio.

Sachion.

Cinchco.

C.Liam

IA.

Meaco.

Scala leucarum; quar

20   40   60   80

# CHINAE,
*olim Sinarum regionis, noua deſcriptio.*
*auctore Ludouico Georgio.*

PAN.

10     15     20     25     30     40     45

In 1514, Johannes Werner, a parish priest in Nuremberg, published a treatise that presented three cordiform, or heart-shaped, projections (one of which was a refinement of a projection developed by Johannes Stabius). These three were equal-area projections, with the distances along each parallel and the central meridian correct – but using them emphasized the northern hemisphere at the expense of the southern.

The work of the Flemish cartographer Gerardus Mercator then began to change the face of map-making. His first triumph came in 1538, when he released a world map using a double-cordiform projection, with one heart for each hemisphere. This was also the first map to record the name "North America".

Mercator's major achievement came in 1569 with a new projection that represented a breakthrough in nautical cartography. Since known as the Mercator projection, it is cylindrical-like, with the meridians as equally spaced, parallel lines and the lines of latitude as parallel, horizontal straight lines, which are spaced farther apart as their distance from the Equator increases. This projection is uniquely suited to navigation because a line of constant true bearing allows a navigator to plot a straight-line course. However, the projection grossly distorts geographical regions in high latitudes – thus, Greenland is shown larger than South America, although it is actually less than one-eighth of the size.

Mercator later began a work that was designed to tell, through maps, the history of the world since its creation. After his death in 1594, editions were produced by Jodocus Hondius, a Flemish cartographer. For the maps of Africa and South America, Hondius used a sinusoidal projection – one with straight-line parallels (like a cylindrical projection), but with curved meridians, all of which are sine curves with the exception of the central one. This projection had first been used for a world map in 1570 by Jehan Cossin of Dieppe.

Another major development in cartography was led by Abraham Ortelius of Antwerp. Already a recognized master, in 1570 he published *Theatrum Orbis Terrarum* (*Theatre of the World*), which is usually considered the first modern geographical atlas. It was a collection of maps of the same dimensions and presented in a uniform style, with a description of each region on the back. Ortelius undoubtedly took much of his information from Mercator, and parts of it are exceptionally accurate. However, as was common at the time, little-known regions have significant errors – including, for example, a non-existent continent in the South Pacific. The first edition contained about 70 maps from the best sources available, but that number increased in later editions, of which there were more than 30.

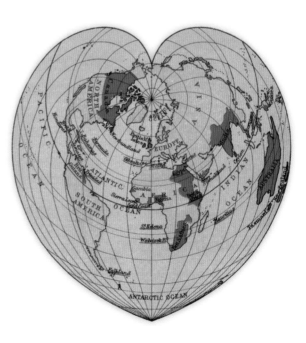

Left: A map of China and the Far East, from an edition of Abraham Ortelius' *Theatrum Orbis Terrarum*, published in Antwerp in 1603. The map is oriented with west to the top, so Bengal is near the top, Borneo to the left (south) and Japan (Iapan) near the bottom (east).

Right: A cordiform (or heart-shaped) projection of the world. This classic projection is actually from the *Harmsworth Universal Atlas and Gazetteer*, published in London in 1908.

# The Dutch and the East Indies

**T**HE PERIOD OF THE LATE SIXTEENTH AND THE
SEVENTEENTH CENTURY is widely recognized as the
Dutch Golden Age, with remarkable achievements
in exploration, mercantilism, art and cartography.
Most of that period overlapped the Low Countries'
struggle for independence from Spain in the Eighty
Years' War (1568–1648) and, although little heralded,
it was a number of changes in the political environment
caused by that war that affected many of the subsequent
developments in exploration and map-making.

The early cartographic centre of the region was
Antwerp, then a major hub of maritime trade and one of
the richest cities in Europe. It was there that Abraham
Ortelius produced his *Theatrum Orbis Terrarum*
(*Theatre of the World*), and where, in 1578, Gerard
de Jode published a major competitive atlas. It was
also in Antwerp that the Flemish cartographer Frans
Hogenberg carried out his early work. However, in the
1570s the region became unstable, and in 1584 Antwerp
was besieged, eventually falling to Spanish troops. Many
of the Protestant citizens moved north to the United
Provinces, an alliance of small duchies and counties that
had declared their independence from Spain in 1581 and
were able to maintain that sovereignty while establishing
a powerful maritime empire. Hogenberg fled from

Opposite: A portrait of Abraham Ortelius with extremely elaborate surroundings, taken from the 1603 edition of his *Theatrum Orbis Terrarum*.

Below: Pages from Lucas Janszoon Waghenaer's *Spieghel der Zeevaerdt*, showing the detail given to the coverage of Europe's coastal waters.

Overleaf: The 1658 edition of *Nova Totius Terrarum Orbis Tabula*, Joan Blaeu's world map that preceded the publication of *Atlas Major*.

Antwerp to Cologne, where he collaborated with Georg Braun, a canon of the cathedral, to produce a six-volume atlas, which included the first systematic collection of plans and images of the world's greatest cities.

Meanwhile, Amsterdam and its environs became Europe's new maritime centre as the Dutch fleets began to dominate commerce both in Europe and in areas further afield previously monopolized by the Spanish or Portuguese. The city also became the focal point of cartographic work: in 1584, Lucas Janszoon Waghenaer, from the nearby port of Enkhuizen, published *Spieghel der Zeevaerdt* (*Mirror of the Sea*), the first printed sea atlas, which covered European coastal waters from Spain to Norway, giving detailed navigational aids, sailing instructions and port information for each of its 45 charts.

Amsterdam also became home to the Flemish cartographer Jodocus Hondius. In the 1590s, he produced a series of world maps, one of which was an early example of cartographic propaganda because its accompanying illustrations were notably anti-Spanish and anti-Roman Catholic. In 1604 Hondius purchased the plates for Mercator's atlas and republished it with 36 additional maps. Hondius, and later his son Hendrick and then his son-in-law Jan Jansson, continued revising and publishing what became known as the Mercator–Hondius Atlas, which carried on for 50 or more editions. The Hondius family was a proponent of the lavish map decoration known as *cartes à figures*, which are ornamental panels depicting individuals, landscapes, and town plans and views, all placed in decorative borders in order not to detract from the map. This was a style later used by many Dutch map-makers.

Like Hondius, the cartographer Pieter van den Keere also had a political agenda. He left Ghent in 1584, when the Spanish regained control, and travelled to London, where his sister married Hondius. After becoming a skilled engraver and cartographer, van den Keere followed Hondius to Amsterdam in 1593. In 1617 he put out *Germania Inferior*, the first "national" atlas of the Low Countries, which included 26 maps and plates of the 17 provinces. One of these maps, entitled "Leo Belgicus", showed the provinces with an outline that made them appear as a lion, which had become the symbol of the United Provinces in the war for independence.

Meanwhile, in 1602, following the success of Dutch merchants in gaining a foothold in the spice trade in Indonesia (a trade that was previously a Portuguese monopoly), the government of the United Provinces granted exclusive regional trading rights to the Dutch East India Company. In the next 60 years, the company took over sites previously held by the Portuguese, established its own plantations, crushed local native opposition and became the largest and richest company in the world, with trading posts as far afield as Java (in what is today Indonesia), Ceylon (Sri Lanka), the Cape of Good Hope, Persia (Iran), Siam (Thailand) and China. In 1621 the Dutch West India Company was founded and given a similar monopoly over the Caribbean, Brazil and North America, as well as the slave trade between Africa and the Americas.

Right: A map of India and Southeast Asia, the region dominated by the Dutch East India Company in the first half of the seventeenth century. The map was produced by Willem Blaeu around 1635, when he was the Dutch East India Company's chief cartographer.

Below: A map of parts the Low Countries, from Pieter van den Keere's *Germania Inferior*, published in Amsterdam in 1617. The atlas' name came from Roman terminology, which referred thus to the low region nearer the sea than Germania Superior. The map is oriented with west to the top.

A natural extension of such a maritime empire was the introduction of new geographic knowledge, although the Dutch East India Company attempted to keep certain discoveries secret. In 1633 the company appointed Willem Blaeu of Amsterdam as its official cartographer. Blaeu's first great cartographic achievement had been the publication in 1608 of a series of sea charts, entitled *Licht der Zeevaert* (*The Light of Navigation*). He had also been working for a long time on an atlas that was finally published in 1635, in two volumes, as *Atlas Novus* (*New Atlas*).

Blaeu died in 1638, but his work was continued by his son Joan, who succeeded him as chief cartographer of the Dutch East India Company. Joan Blaeu's greatest achievement was *Atlas Major*, the publication of which was completed in Latin in 11 volumes in 1662, followed by Dutch, German, French and Spanish editions. The mammoth work contained nearly 600 maps and about 3,000 pages of text. Although its geographical content was not as accurate as later works, it was arguably the most opulent atlas ever and is generally regarded as the pinnacle of atlas publishing.

Below: A map view of Jerusalem from the six-volume *Civitates Orbis Terrarum*, published in Cologne between 1572 and 1617. Most of the maps and prospects of the world's cities were edited by Georg Braun and engraved by Franz Hogenberg.

# French and English
## Mapping in North America

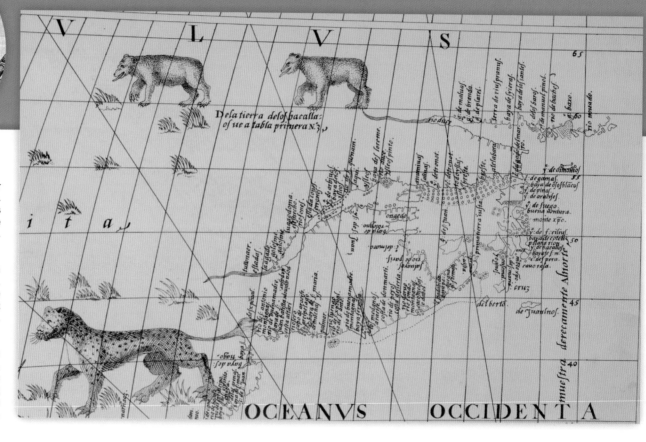

Above: John White's watercolour "Indians fishing", painted in 1585–86. It shows several ways the inhabitants caught fish, and a broad range of the sea creatures they hunted.

Right: Detail of eastern Canada from Sebastian Cabot's world map of 1544. Human and animal figures dotted the interiors of those continents not yet explored by Europeans.

Below: A replica of John Cabot's ship *Matthew* in Bristol, England. In 1997, the new *Matthew* followed Cabot's course to Newfoundland, carrying the same number of crew for the same period of time.

ALTHOUGH LITTLE IS CERTAIN ABOUT THE ORIGINS OF JOHN CABOT, he was officially of Venetian nationality when he received letters patent from King Henry VII of England (reigned 1485–1509) to sail west from Bristol to reach the Indies. In June 1497, having taken a more northerly route than Columbus, he made landfall on the coast of North America, possibly on what is today Cape Breton Island or Newfoundland. His voyage brought North America to the attention of rulers, merchants, explorers and cartographers. The first map known to show Cabot's voyage was that produced by Juan de la Cosa shortly after 1500.

Cabot's son Sebastian succeeded him as an explorer, commanding expeditions to both North and South America, the latter for Spain. Around 1544, Sebastian Cabot produced a world map that later disappeared for several hundred years, until it was rediscovered in 1843 at the home of a Bavarian curate.

Around the same time that Sebastian Cabot's world map was published, Jean Rotz, a Franco-Scottish cartographer, finished his *Boke of Idrography* (*Book of Hydrography*), which he had intended to present to King Francis I of France (reigned 1515–47), but instead dedicated to England's King Henry VIII (reigned 1509–47). Rich with flowery sidebars, colourful compass roses and imagined scenes of various lands – all aspects of what became known as the Dieppe school of cartography – the book included a map of the world and 11 regional nautical charts, including one of the North Atlantic, which showed the coastlines of what are now Newfoundland and Labrador.

Meanwhile, the interior of North America received proportionately little attention from map-makers, despite expeditions such as those led by Juan Ponce de León to Florida (1513), Hernando de Soto, in what became the southeastern United States (1539–43), and Francisco de Coronado, who searched for the "Seven Cities of Cibola" in the American west (1540–42). During his three expeditions to the Gulf of St Lawrence and its river system (1534,

1535–36 and 1541–43), French explorer Jacques Cartier reputedly made maps of eastern Canada, but, according to the great chronicler of exploration Richard Hakluyt, the maps were then lost.

On the western side of the continent, California received cartographic attention after Sir Francis Drake's circumnavigation of the Earth (1577–80). During this journey Drake's *Golden Hind* passed through the Strait of Magellan at the tip of South America before sailing up the west coast of the Americas as far north as San Francisco. He then set off across the Pacific Ocean. His return to England coincided with the period in which Jodocus Hondius lived in London after fleeing Flanders. Hondius produced multiple portraits of the celebrated mariner. After moving to Amsterdam in 1593, Hondius published several world maps, one of which displayed the routes of both Drake and Thomas Cavendish, who had followed Drake around the world in 1586–88.

Below: A nautical chart of the region surrounding the Caribbean Sea, from the *Boke of Idrography* by Jean Rotz. The map is oriented with south to the top, which is the northern coast of South America. The islands of the Caribbean are in the centre and the east coast of North America below that, extending from Florida to the region of Newfoundland and the Gulf of St Lawrence.

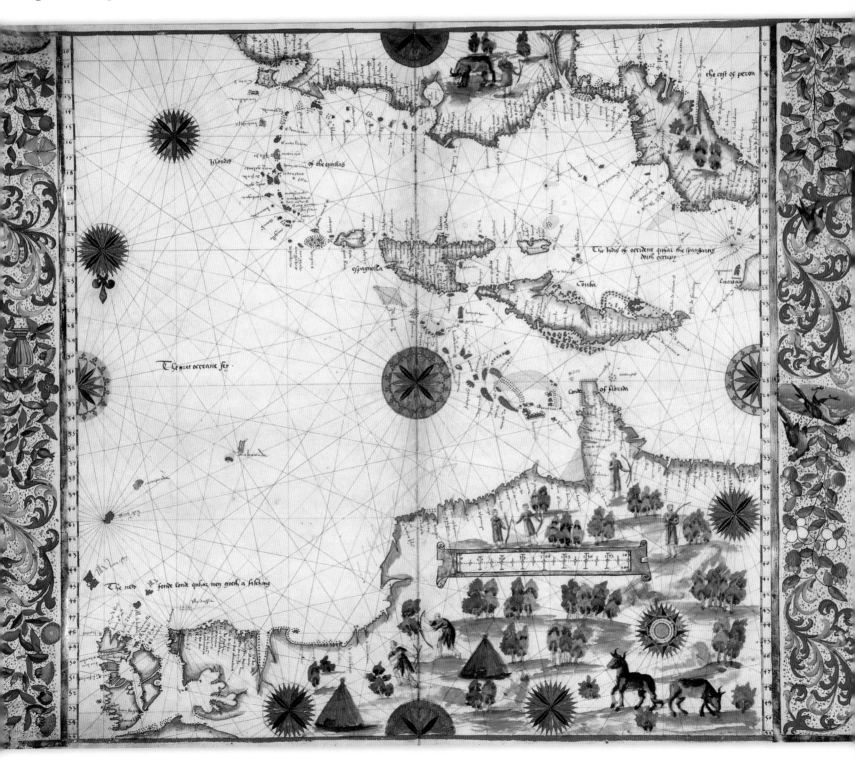

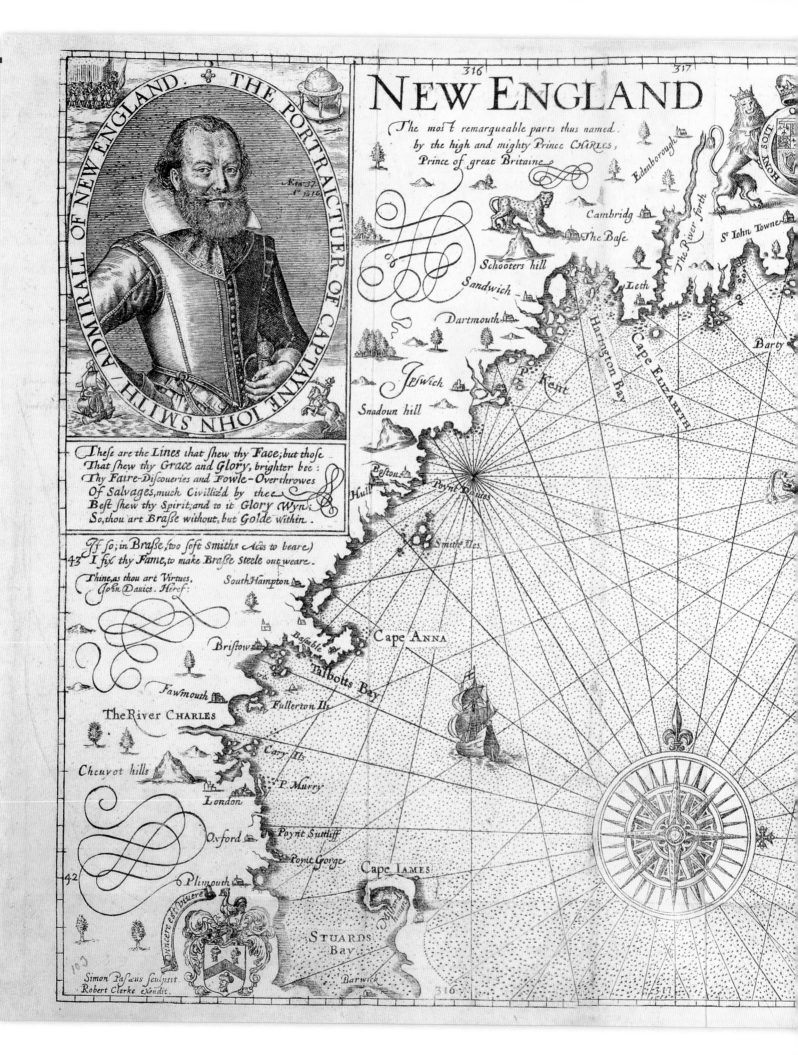

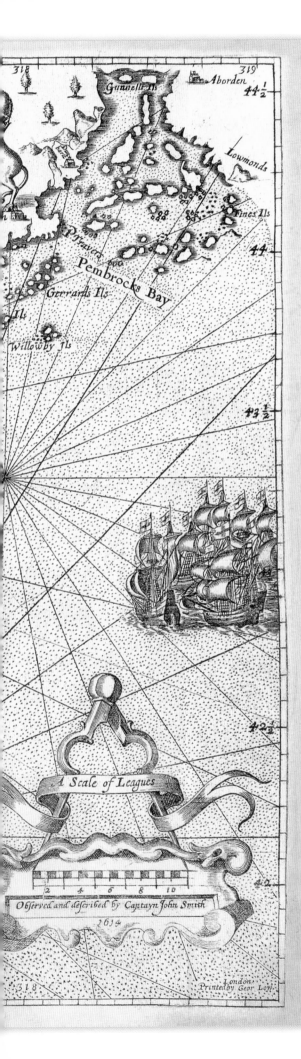

Towards the end of the sixteenth century, accurate regional maps of North America began to be produced. John White was a talented artist who sailed to Roanoke Island off the coast of North Carolina three times in the 1580s to help establish a colony envisaged by Sir Walter Raleigh. White produced a series of watercolours of the local native peoples and landscape, and he and Thomas Harriot made surveys in the vicinity of Roanoke Island, as well as of the coastline from Florida to the Outer Banks of North Carolina. White's charts were the earliest known efforts by the English to map their new possessions in North America.

Perhaps the most impressive regional map of the New World was one compiled by Samuel de Champlain during the first five of his seven expeditions of discovery to New France (present-day Canada) between 1603 and 1611. Unlike most previous charts, which were drawn by cartographers using information obtained from explorers and navigators, this portolan-style chart on vellum was produced entirely by Champlain, based on his own observations and calculations, as well as his own interviews with locals. The map provided the first thorough delineation of the coast of what became Canada and New England, and the coastline and place names closely corresponded with his 1613 expedition account, giving even more accurate detail to the image that he was creating of New France.

Soon after Champlain had completed his map, an English chart appeared that covered some of the same region. John Smith had been a member of the Jamestown settlement in Virginia in 1606–09, during which time he had explored and surveyed the Rappahannock river and Chesapeake Bay. In 1612, he published a gloriously illustrated map of the region, including the village where, in a famous incident, he had been saved from execution by Pocahontas, daughter of the Indian chief Powhatan.

In 1614 Smith explored and surveyed the coast from Cape Cod to Maine. He later tried several times without success to establish settlements in New England, a name he introduced to popular usage. Captured by French pirates in 1615, Smith used his time in captivity to produce a detailed map of the coastline of the American northeast, which was subsequently used by the Pilgrim Fathers. The map was published with Smith's book *A Description of New England* in 1616, complete with names transplanted from Britain, thereby showing the "Englishness" of the region, which he hoped would encourage others to settle there.

Left: John Smith's map of the coastal regions of New England. The map first appeared in 1616 in Smith's book *A Description of New England*. This engraving was produced for his publication of 1624, *The Generall Historie of Virginia, New-England, and the Summer Isles*.

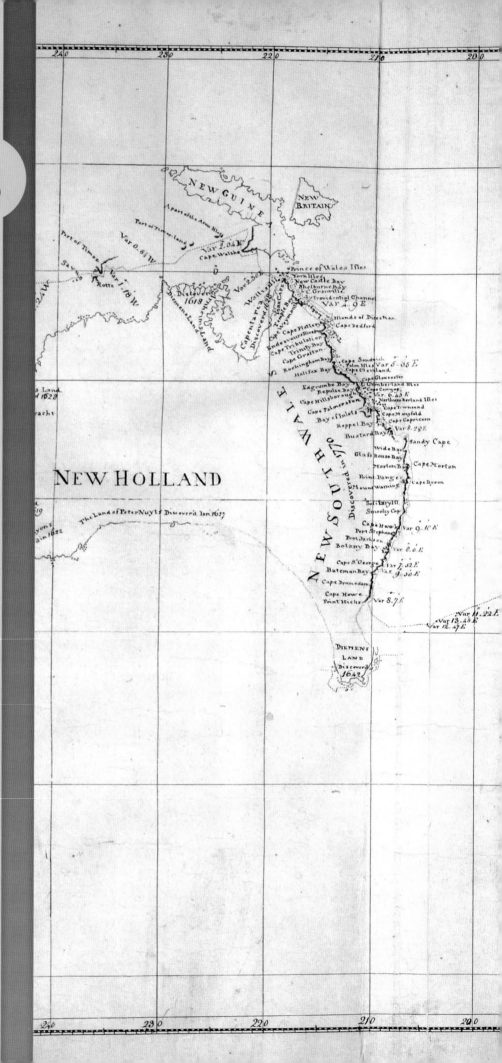

# CHAPTER **3**

# The *World* *Expands*
## – Filling in the Gaps
## (1600–1800)

Right: To the eighteenth-century public in Britain, the details on this map – A Chart of the Great South Sea or Pacifick Ocean – would have been as new as if they had been of the Moon. Produced by James Cook and showing the route of his ship *Endeavour* in the middle years of his first expedition (1769–70), it featured previously little-known locations such as the Society Islands, New Zealand and New South Wales.

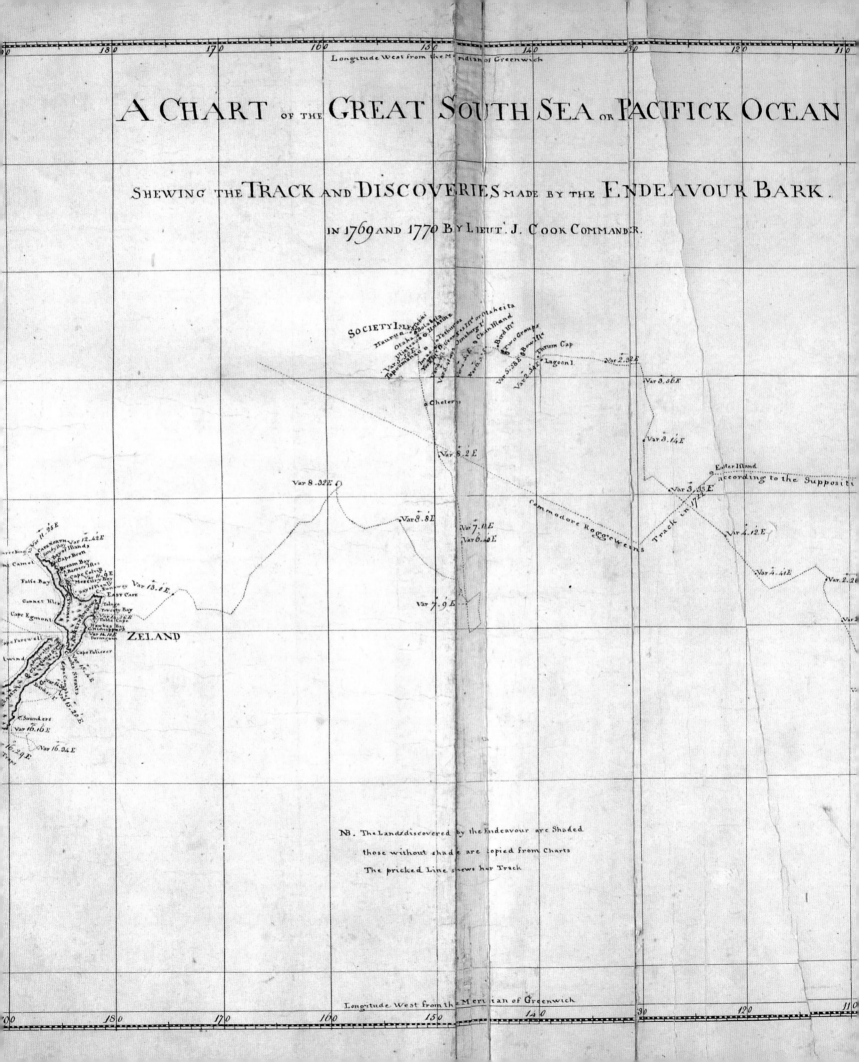

# A CHART OF THE GREAT SOUTH SEA OR PACIFICK OCEAN

## SHEWING THE TRACK AND DISCOVERIES MADE BY THE ENDEAVOUR BARK.

### IN 1769 AND 1770 BY LIEUT. J. COOK COMMANDER.

SOCIETY Isles

ZELAND

Var 2.32 E

Var 3.56 E

Var 3.14 E

Easter Island according to the Suppositi

Var 8.2 E

Var 3.33 E

Var 8.32 E

Commodore Roggewens Track in 1722

Var 8.8 E

Var 4.12 E

Var 7.12 E

Var 6.40 E

Var 4.41 E

Var 2.2

Var 7.9 E

NB. The Lands discovered by the Endeavour are Shaded

those without shade are copied from Charts

The pricked Line shews her Track

# National Mapping

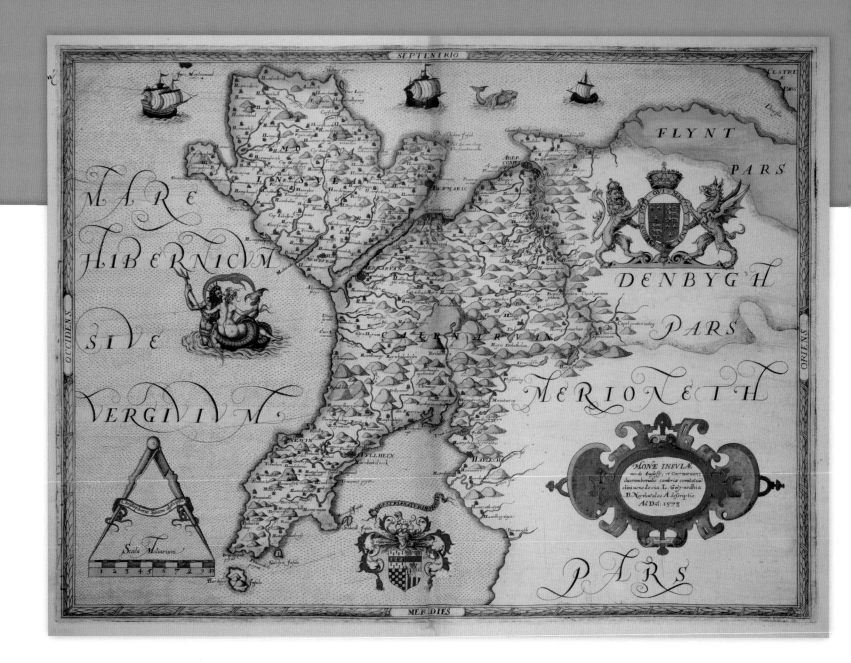

**N**OT LONG AFTER THE RULERS, administrators of mercantile companies, explorers and colonial settlers of far-flung reaches of the world began to understand the increasing value of geographically accurate maps of those places, the ministers, estate-holders and tax-collectors in the homelands of Europe also started to see their worth. However, by the beginning of the seventeenth century, local and national maps were produced for purposes other than to satisfy officials. A significant proportion of the population in many European countries was literate and had capital to spend on luxuries, which meant that printed

maps, whether they were wall-decorations or books, became big business for cartographers and printers.

One of the more glorious and accurate examples of an early regional map is the one known as the Carta Marina, which has a highly unusual provenance in that it is a map of Scandinavia printed by the renowned woodcut artists and printers of Venice. The Carta Marina was produced by Olaus Magnus, who travelled to Rome in 1537 from his native Sweden with his brother Johannes, who, as the Roman Catholic archbishop of Uppsala, was effectively going into exile

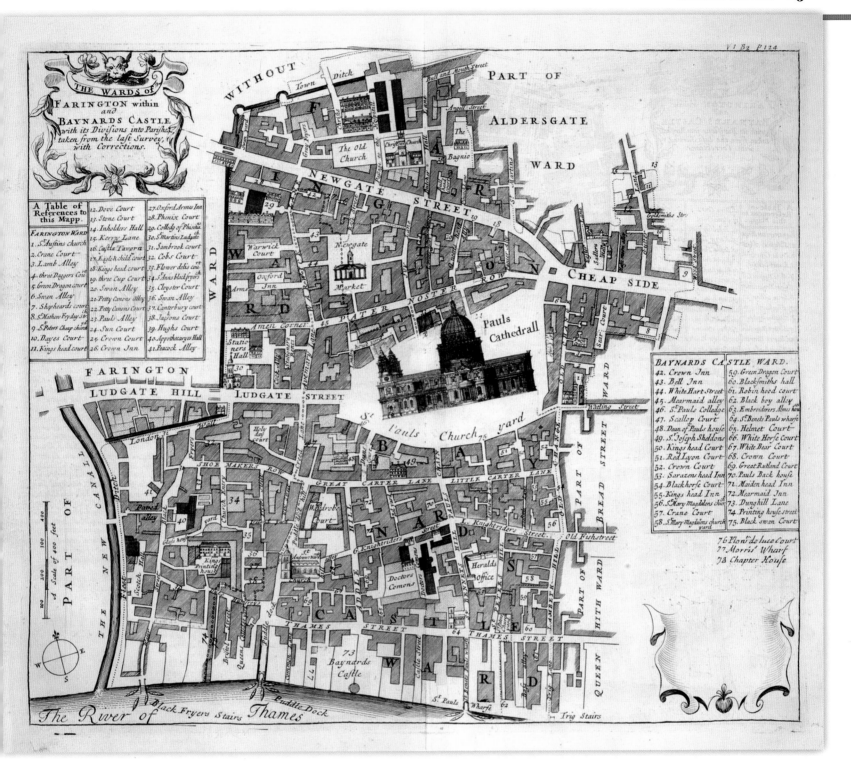

VI B2 P124

THE WARDS of
FARINGTON within
and
BAYNARDS CASTLE
with its Divisions into Parishes,
taken from the last Survey,
with Corrections.

A Table of
References to
this Mapp.

FARINGTON WARD
1. St Austins church
2. Crane Court
3. Lamb Alley
4. three Daggers Cou
5. Green Dragon court
6. Swan Alley
7. Shepheards court
8. St Mathew Fryday str
9. St Peters Cheap churc
10. Dayes Court
11. Kings head court

12. Dove Court
13. Stone Court
14. Inholders Hall
15. Kerry Lane
16. Castle Tavern
17. Eagle & child cour
18. Kings head court
19. three Cup court
20. Swan Alley
21. Petty Canons Alley
22. Petty Canons Cou
23. Pauls Alley
24. Sun Court
25. Crown Court
26. Crown Inn

27. Oxford Armes Inn
28. Phenix Court
29. Colledg of Phisitia
30. St Martins Ludgate
31. Sambrook court
32. Cobs Court
33. Flower delis cou
34. St Anns black fryers
35. Cloyster Court
36. Swan Alley
37. Canterbury court
38. Jacksons Court
39. Hughs Court
40. Appothecaryes Hall
41. Peacock Alley

BAYNARDS CASTLE WARD.
42. Crown Inn
43. Bell Inn
44. White Hart Street
45. Mearmaid alley
46. St Pauls Colledge
47. Scallop Court
48. Dean of Pauls house
49. St Joseph Sheldons
50. Kings head Court
51. Red Lyon Court
52. Crown Court
53. Sarazens head Inn
54. Blackhorse Court
55. Kings head Inn
56. St Mary Magdalens chu
57. Crane Court
58. St Mary Magdalens church
yard.

59. Green Dragon Court
60. Blacksmiths hall
61. Robin hood court
62. Black boy alley
63. Embroiderers Almes hou
64. St Benets Pauls wharfe
65. Helmet Court
66. White Horse Court
67. White Bear Court
68. Crown Court
69. Great Rutland Court
70. Pauls Back house
71. Maiden head Inn
72. Mearmaid Inn
73. Dunghill Lane
74. Printing house street
75. Black swan Court

76. Flour'de luce Court
77. Morris' Wharf
78. Chapter House

because Sweden had officially adopted Lutheranism. Geographically superior to most maps of its time, it was intensively illustrated with native peoples, wildlife and sea monsters. It first appeared in 1539, predating by almost two decades Magnus's monumental work on the history and culture of the north, *Historia de Gentibus Septentrionalibus* (*History of the Northern Peoples*). The map was long thought lost, until it was discovered in Munich in 1886.

Opposite: Christopher Saxton's map of northwest Wales and the island of Anglesey from his masterpiece *An Atlas of England and Wales*, published in 1579.

Left: Photo of the Observatoire de Paris. Inside the building is the Meridian Room (or Cassini Room), in which the Paris meridian is traced on the floor.

Above: A map of the Ward of Farington, including St Paul's, from John Strype's 1720 expanded edition of John Stow's *Survey of London*

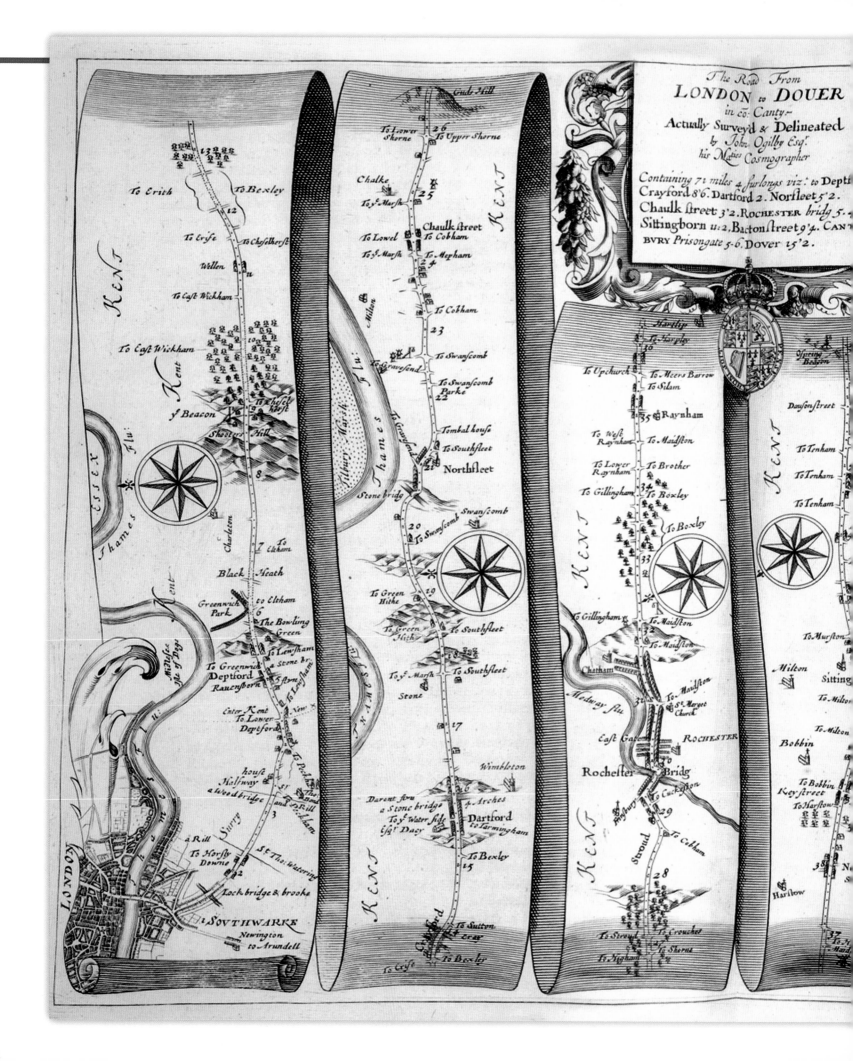

The Road From
LONDON to DOUER
in co. Cant:—
Actually Surveyd & Delineated
by John Ogilby Esqr.
his Maties Cosmographer

Containing 71 miles 4 furlongs viz: to Deptf
Crayford 8'6. Dartford 2. Norfleet 5'2.
Chaulk street 3'2. Rochester bridg 5.
Sittingborn 11:2. Bacton street 9'4. CANT
BVRY Prisongate 5:6. Dover 15'2.

**First strip (London to Southwarke):**

To Erith    To Bexley
12
To Crise    To Chesclherst
Wellen    11
To East Wickham
To East Wickham    10
To Chiselhurst
9
y.e Beacon
Shooters Hill    8
Charlton
7    To Eltham
Black Heath
Greenwich Park    to Eltham
6    The Bowling Green
To Lewsham
To Greenwich    a Stone br.
Deptford    To Lewsham
Ravensborn    5    Flyrn
Enter Kent    Vent x
To Lower Deptford
house    Halfway    To Peckha
a Wood bridge    4    Tho: Nob.
and yt Rill
3
a Rill    Surry
To Horsly Downe    St. Tho: Watering
2
LONDON    Lock bridge & brooke
SOVTHWARKE
Newington    to Arundell
KENT
ESSEX Flu:
Thames
Midlesex Isle of Dogs
Surry Flu:

**Second strip (Southwarke area to Dartford):**

Guds Hill
To Lower Shorne    26    To Upper Shorne
Chalke
To ye Marsh    25
Chaulk street
To Lowel    To Cobham
To ye Marsh    To Mepham
24
To Cobham
23
To Swanscomb
To Swanscomb Parke    22
Tombal house
To Southfleet
21    Northfleet
Stone bridg
20    Swanscomb
To Swanscomb
To Green Hithe    19
To Green Hithe    To Southfleet
18    To Southfleet
Stone
Wimbloton
Darent flow    a Stone bridge    4 Arches
To ye Water side    Dartford
Esq. Dacy    to Farmingham
To Bexley
15
To Cr[i]se    To Sutton at [...]
To Bexley
KENT
Thames Flu:
Tilbury Marsh
Milton

**Third strip (Gravesend to Rochester):**

Hartlip
To Hasley    30
To Upchurch    To Meers Barrow
To Silam
35    Raynham
To West Raynham    To Maidston
To Lower Raynham    To Brother
34    To Boxley
To Gillingham
To Boxley
33
To Gillingham    To Maidston
32    To Maidston
Chatham
Medway flu
East Gate    St. Margret Church
To Maidston
ROCHESTER
Rochester    Bridg
Cuckston
Frinbury    29
Stroud    To Cobham
28
To Stroud    To Crouchet
To Higham    To Shorne
KENT

**Fourth strip (Sittingbourne area):**

Ospring Boxton
Dawson street
To Tenham
To Tenham
To Tenham
To Murston
Milton    Sitting[born]
To Milton
To Milton
Bobbin
To Bobbin
Key street
To Harston
36
35
Harston    37
KENT

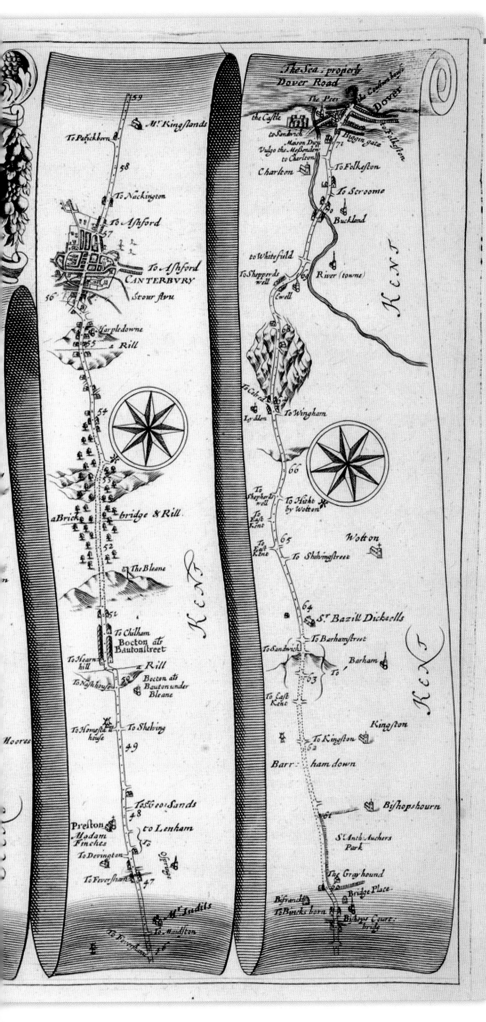

Several decades after Magnus, Philipp Apian, the son of Petrus Apianus, was a professor of mathematics and astronomy at the University of Ingolstadt when he was commissioned by the duke of Bavaria to produce a map of the region. After seven years of surveying, in 1561–63 Apian created a map of Bavaria that measured 5 × 5 m (16 × 16 ft), the accuracy of which remained unsurpassed until it was destroyed by fire in 1782.

In Europe, printers were quick to see the financial advantages of producing local and national maps. That was not the case in England, where it was not until the early 1570s that Christopher Saxton was commissioned to survey the kingdom by William Cecil, Lord Burghley, the secretary of state for Queen Elizabeth I (reigned 1558–1603). Saxton divided his maps into counties, completing Norfolk first in 1574. Four years later he had finished, and in 1579 an atlas containing 34 county maps, one map of England and Wales, and a list of cities and towns was published. However, the maps lacked standardization and the sheet sizes were different, as were some of the symbols and occasionally the orientation. Nevertheless, apart from Apian's maps, the undertaking was unparalleled in Europe for its organization and quality.

The success of Saxton's work meant that it was immediately followed by competing sets of county maps, including one by John Norden. However, the best compilation was one by John Speed, which was published in 1611 as *Theatre of the Empire of Great Britain*. Speed's county maps offered extensive decoration and information, including plans of major towns, the coats of arms of a particular county's leading families, images of key political figures, illustrations of important sites such as castles, and extensive geographical and political detail. Each map could have colour applied to it if the purchaser desired. The engraving of the maps was carried out by former London resident Jodocus Hondius.

For much of the seventeenth century, most of the maps produced in Britain were reworkings of the earlier efforts, and it was not until 1675, with the publication of John Ogilby's *Britannia*, that the next true step forward in English cartography occurred. Ogilby's masterpiece included 100 pages of maps and 200 pages of additional text covering 73 postal roads of England and Wales, produced in the form of continuous parallel strips, side by side. Each strip contained a wealth of information relevant to that particular road, such as hills and rivers, nearby estates and churches, indications of the type of bridges, and whether parts of the road were open or enclosed by hedges. Approximately 12,000 km (7,500 miles) of roads were charted, at a uniform scale of 2.5 cm per 1.6 km (1 in per statute mile). Sadly, Ogilby died a few months after the book was released, and it took more than 40 years for a portable edition to be produced.

Left: The six strips illustrating the road for those travelling from London to Dover, as shown in John Ogilby's *Britannia*, published in 1675.

## Cassini and Newton

**When Jean-Dominique Cassini completed his measurement of the arc of the meridian, he concluded that Isaac Newton's 1687 description of the Earth as an oblate spheroid somewhat flattened at its poles was incorrect. Cassini instead agreed with the Cartesian view that the Earth is somewhat elongated, with a polar diameter greater than its equatorial one.**

**In 1720, after the completion of the triangulation between Dunkirk and Perpignan, Jacques Cassini supported his father's view. However, in the following years, a growing body of evidence – including the meridian measurements by Pierre Maupertuis's expedition to Lapland, and the French Geodesic Mission to Quito, at the Equator, in what is now Ecuador – supported Newton. After earlier backing the position taken by his father and grandfather, César-François Cassini, accepted that Newton's description was correct.**

Above: Sir Isaac Newton (1643–1727), whose genius and influence extended to physics, mathematics, astronomy, optics and natural philosophy. The portrait was painted of him at the age of 59.

Around the time that the new edition of Ogilby appeared, so too did another reprint, of a much earlier work. In 1598 John Stow had published his *Survey of London*, a descriptive and anecdotal account of the buildings, streets, inhabitants, social conditions and customs of the Tudor capital. In 1720 the ecclesiastical historian John Strype published an expanded version of Stow's work, which updated information about the rebuilding of the city after the Great Fire of 1666 and its subsequent growth into a sprawling metropolis. Equally significant, the new edition added a stunning collection of ward and parish maps that made London the most thoroughly and accurately mapped city in the world.

While Ogilby had been producing *Britannia*, across the English Channel extraordinary work was being undertaken that would have a permanent influence on cartographic endeavours. In 1669–70, Jean Picard, the prior of Rillé in Anjou and a professor of astronomy at the Collège de France, carried out a triangulation survey in order to make a measurement of an arc of the meridian (a line of longitude). Using a chain of 13 triangles, he measured one degree of longitude from the clock tower in Sourdon, near Amiens, to near Paris, and from that he was able to make the first accurate calculation of the circumference of the Earth. Once finished with that project, Picard extended his triangulation survey all the way to the English Channel, which initiated the first national surveying and mapping programme in history.

Above: An engraving of Jean Picard (1620–82), the French priest and astronomer whose measurement of an arc of the meridian initiated the first national surveying and cartographic programme.

In 1683, Jean-Dominique Cassini, the Genoa-born director of the Observatoire de Paris (Paris Observatory), began to extend the survey that Picard had begun. For almost 30 years he surveyed the Paris meridian from Dunkirk to Perpignan, working alongside his son Jacques, who eventually completed the work in 1718, having succeeded his late father as director of the observatory in 1712.

Meanwhile, Guillaume Delisle, a student of the elder Cassini, began producing the first maps using Cassini's concept of fixing accurate positions by astronomical observation. Many map-makers did not understand how to turn these "theoretical" data from Cassini's surveys into usable measurements. Delisle successfully incorporated this astronomical and mathematical scientific material, while at the same time introducing more rigorous requirements for the use and analysis of source and ground material. The consequent reduction of errors and the accuracy of his charting made his maps far superior to earlier ones. Delisle therefore not only produced the most accurate maps of France yet published, but he also turned out the most detailed maps of New France.

Such was Delisle's reputation that Tsar Peter I (Peter the Great, reigned 1682–1725) invited his younger brother, Joseph-Nicolas Delisle, to establish an observatory in St Petersburg and conduct a geodetic survey of Russia. In the next two decades, Joseph-Nicolas played a major – and unfortunate – role in Russian cartography and exploration. Vitus Bering's first expedition to Siberia (1725–30) had indicated that Asia and North America were not connected. In 1734, Ivan Kirilov, who had directed a topographical survey of Russia, began the publication of his impressively accurate *Atlas of Imperial Russia*. At the same time, the Russian Admiralty asked the country's Academy of Sciences to produce a map of the North Pacific for use on Bering's next expedition. The Academy of Sciences commissioned Delisle for this task.

Delisle produced a fanciful – incorporating several apocryphal stories from the seventeenth century – and, in places, wildly inaccurate map of a virtually continuous Siberia and North America, with a large landmass off the coast of Asia called Juan da Gama's Land. Unfortunately, Kirilov died before his atlas could be completed, and the

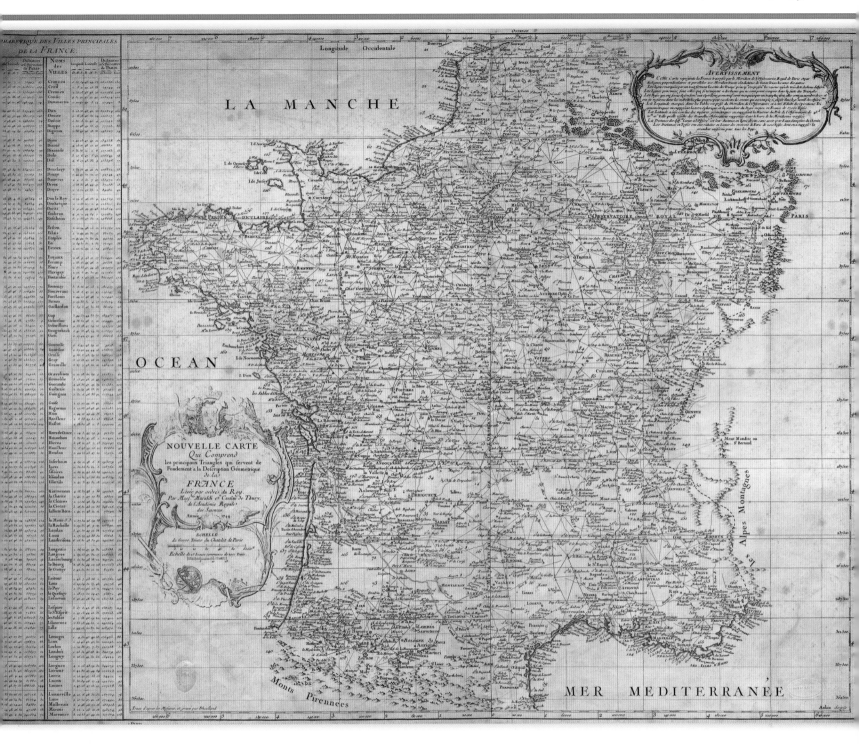

Above: The magnificent 1744 map of France produced by the Cassini family after the completion of the first national mapping survey.

Russian government, due to political pressure from the Academy of Sciences, ordered Bering to follow Delisle's map rather than that of Kirilov, which the academy attempted to suppress. Consequently, much time was lost on Bering's expedition.

Throughout this period, the Cassini family expanded its triangulation to include all of France. Jacques Cassini and his son, César-François, began to prepare the first set of maps based on a complete triangulation and topographic survey of an entire country. In 1744, the 18-sheet map of France was published, but it did not take long for the Cassini dynasty to expand its project.

Within three years King Louis XV (reigned 1715–74) had requested that César-François embark on a larger set of maps, although government support was later withdrawn due to lack of funds. César-François formed a company (later run by his son, Jacques-Dominic) to undertake the project, and the two oversaw the work of some 80 surveyors and cartographers for the next five decades. The production of the final maps – all 181 of them, in what came to be known as the *Carte de Cassini* – was interrupted by the French Revolution, but they continued to appear thereafter, with the final ones issued in 1815.

# Men, Measurements and Mechanisms

**T**HE CONTINUING IMPROVEMENTS in the geographical accuracy of maps throughout the seventeenth and eighteenth centuries were not just the results of cartographers, engravers and printers refining their crafts. Astronomers, navigators, mathematicians and inventors all contributed to advances in the techniques and instruments of measurement, which combined to make land maps and maritime charts more precise than ever before.

One of the giants in this process was Gemma Frisius, the mentor of Mercator and a professor of medicine at the Catholic University of Leuven in the Low Countries, who was also a theoretical mathematician and instrument-maker. In 1533 Frisius made the first-known proposal for using triangulation as an accurate method for locating distant places. He also improved a number of astronomical and navigational tools, such as the cross staff. This was used aboard ships to measure the angle of the Sun above the horizon at noon, which, when used in conjunction with navigation tables, allowed a determination of latitude. Frisius's new cross staff included brass sights and sliding vanes, which spared the observer from having to look directly into the Sun – instead, he could align the horizon with the Sun's shadow on a vane at the end of staff, thus allowing the angle to be determined. Frisius also improved the astrolabe and astronomical rings; the latter was, essentially, a small, portable armillary sphere to show the motions of the heavens. Shortly before he died, aged just 47, he theorized on how an accurate clock might be used to determine longitude.

Around the period that Frisius flourished, the plane table was also developed, which provided a solid, level surface for surveying. The first-known mention of a plane table came in 1551, and an early description also appeared in *Pantometria*, a 1571 book published by Thomas Digges, but based on notes and observations made by his father, Leonard. The elder Digges has also been credited with the invention of both

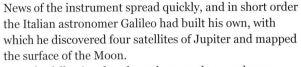

Opposite top left: An engraving of Gemma Frisius holding a set of astronomical rings, also known as Gemma's rings. Frisius updated ancient instruments into what was essentially a simplified, portable armillary sphere.

Opposite below right: An observation being taken by means of a large azimuth quadrant. Initially developed by the German mathematician and astronomer Peter Cruger, this quadrant's design was later completed by Johannes Hevelius.

Below: A painting of the famous astronomer Galileo Galilei in a contemplative mood. The picture was painted by Ivan Petrovich Koler-Viliandi in 1858.

the theodolite – a tool used to measure horizontal and vertical angles – and the telescope, although the latter is disputed. Digges certainly experimented with lenses and mirrors, although they were not of a very advanced design.

The theodolite mentioned in *Pantometria* was actually an azimuth instrument, which only measured horizontal angles. Confusingly, Digges also described what he called a "topographicall instrument", which could measure both vertical and horizontal angles, and was later sometimes called an altazimuth; today, this is the accepted definition of a theodolite. The first true theodolite is generally credited to the German Joshua Habermel, who produced it complete with compass and tripod in 1576.

Although Digges might have envisioned a telescope, the three earliest known examples were produced independently around 1608 in the United Provinces by Hans Lipperhey, Zacharias Janssen and Jacob Metius.

News of the instrument spread quickly, and in short order the Italian astronomer Galileo had built his own, with which he discovered four satellites of Jupiter and mapped the surface of the Moon.

In the following decades, telescopes became larger, better focused and more powerful, enabling astronomers to make more discoveries. Two cartographers of the heavens were particularly notable, Michel Florent van Langren and Johannes Hevelius. In 1645, van Langren made the first map of the Moon, as seen through a telescope, and he assigned names to many features. Concurrently, Hevelius built an observatory on top of three neighbouring houses in his native Danzig (modern-day Gdańsk, Poland), where he placed his 18-m (60-ft) telescope. Hevelius wrote *Selenographia*, the first atlas of the Moon, which was published in 1647. He also made a catalogue of 1,564 stars, charted several constellations for the first time and constructed a 45-m (150-ft) Keplerian telescope (one with a convex lens as the eyepiece rather than a concave one, as used by Galileo), which was probably the longest in the world at the time.

In 1615, while Galileo was scanning the skies, Willebrord Snell of Leiden conducted the first major triangulation. It had long been accepted that if the measurements of one side and two angles of a triangle were known, the other two sides and the angle could be calculated. Therefore, beginning with a known baseline, Snell measured the angles from the baseline's endpoints to a distant landmark. He then used these distances as new baselines and constructed a series of triangles until he had a network of 33, which extended approximately 111 km (69 miles) from Alkmaar to Bergen op Zoom. As these Dutch towns were separated by precisely one degree on the meridian, he was able to arrive at a figure for the circumference of the Earth.

Some 50 years later, Jean Picard followed with his own meridian survey, using improved instruments. One of these was a terrestrial telescope that had cross wires placed at the common focal point of the lenses to define the centre of the field of view, thus serving as a telescopic sight. This was attached to a quadrant, and with the additional use of a micrometer – originally developed in the 1630s by English astronomer William Gascoigne, along with the telescopic sight – Picard's work acquired a previously unattainable degree of accuracy.

Picard has been credited with the development of a reflecting quadrant, as have Robert Hooke, Edmond Halley and Isaac Newton. All were early examples of instruments using mirrors to allow the measurement of the altitude of an astronomical object or the angular distance between two horizontal objects by observing them simultaneously. This facility meant these devices had applications for maritime navigation and terrestrial surveying as well as astronomy.

Right: An early example of an octant, which could be used both for surveying and navigation. This model was designed and developed by John Hadley in 1730.

Below: A sextant used by explorer David Livingstone in central Africa in the 1860s. A simple comparison shows the evolution from the octant to this sextant.

The next step in the development of an observing instrument came in around 1730, when John Hadley of London and Thomas Godfrey of Philadelphia independently developed an octant. Named because its arc was one-eighth of a circle, an octant's mirrors allowed measurements of angles up to 90°. For example, to measure the angle of a celestial object, an observer would focus directly on the horizon, while a mirror pivoting on the index arm provided a reflected image of the object. When this was aligned, the observer read the angle from a scale at the other end of the index arm. Thereafter it was easy to calculate the latitude.

In the 1750s Admiral John Campbell of the Royal Navy recommended that an instrument be developed to increase the octant's potential angle of measurement. In response, John Bird created a sextant that could measure to angles of 120°. (In later years, the sextant was refined for specific and differing aspects of navigation and surveying.)

However, one major problem continued to face maritime surveying: sailors had no accurate or reliable way of finding longitude at sea. They could rely on dead reckoning, but this was inaccurate on long voyages, and on occasion had resulted in disaster. Mariners could measure the local time by observing the Sun, and if they knew the time at a

fixed location, the difference between that and local time meant that they could calculate their longitude. But how could time be determined at a distant reference point while aboard ship?

From the sixteenth century onwards, there had been efforts to resolve this question. In 1567, King Phillip II of Spain had offered a prize, and King Philip III later increased it. In 1636, the United Provinces followed suit, and in France the Académie des Sciences (Academy of Sciences) thereafter. In 1714, Parliament in the United Kingdom promised £20,000 to anyone who could solve the problem and a government body popularly known as the Board of Longitude was established to oversee the efforts.

Not that financial encouragement was needed. Beginning in 1514 with Johannes Werner, numerous astronomers had proposed astronomical methods – their ranks included Galileo, Jean-Dominique Cassini, Edmond Halley, Tobian Mayer and John Flamsteed, the first British Astronomer Royal. Most of these suggestions revolved around the "lunar distance method", the notion of compiling an accurate catalogue of the position of the stars, against which the motion of the Moon could be measured, with tables of the Moon's relative position being used to calculate the time at a set location.

Left: H4, the marine chronometer that looked like a large pocket watch. Designed by John Harrison, it finally resolved the "longitude problem" that had plagued mariners for centuries.

Below: An engraving of John Harrison, shown with H4. Even though Harrison designed a successful chronometer, the Board of Longitude made him wait years for all of his prize money.

Another solution – proposed by Frisius – was to carry aboard the ship a timepiece that would maintain correct time regardless of the weather and the sea conditions. Although accurate pendulum clocks existed in the seventeenth century, the rolling and pitching of a ship, the physical pounding the vessel received from the waves, constant dampness, and the changes in humidity and temperature all prevented such clocks from keeping accurate time at sea.

The first serious attempt to produce what became known as a marine chronometer was made in 1657 by Christiaan Huygens, a Dutch physicist, who invented the pendulum clock. His endeavour was based in part on the balance spring watch – which he developed separately from, but concurrently with, Robert Hooke – but despite Huygens's efforts the clock remained imprecise at sea.

The longitude problem was eventually solved by the Englishman John Harrison. After five years of preparation, he tested his first chronometer in 1736 on a voyage to Lisbon, which revealed inadequacies that encouraged him to build a second. During the next 22 years, he built two more chronometers, but these did not pass the Board of Longitude's tests, so he opted for a totally new design. He solved the problem during the next four years by building a chronometer that looked like a large pocket-watch. H4, as it was known, was only 13 cm (5 in) in diameter and weighed just under 1.5 kg (3.3 lb). In 1761–62, the chronometer was tested on a voyage to Jamaica and found to have lost only 5.1 seconds.

Despite this incredible success, the Board of Longitude, led by Astronomer Royal Nevil Maskelyne, remained sceptical and withheld parts of the prize for many years. Nevertheless, Harrison had changed the face of maritime travel and nautical charting forever. In 1772, on his second great voyage of discovery and survey, Captain James Cook took with him a duplicate of H4, which he praised fulsomely on his return.

# Mapping Australia and the Pacific

Right: A painting of Captain James Cook, the great maritime explorer whose investigations ranged from the Arctic to the Antarctic and included vast swathes of the Pacific Ocean.

I N 1605, THE DIRECTORS OF the Dutch East India Company sent a message to the governor of Batavia (present-day Jakarta), the centre of Dutch commercial operations in the East Indies: "There must be more charting, mapping and exploring of the lands farther east of the Spice Islands," they ordered, "and a renewed search for a passage through to the Pacific Ocean."

In response, an expedition sailed the next year. Willem Janszoon, commander of the tiny ship *Duyfken*, followed the south coast of New Guinea, turned south, spied land and came ashore on the west coast of what is today Cape York Peninsula in northeast Australia. The men were the first Europeans to land in Australia. Soon they would be followed by more Dutch ships: for example, in 1616, while sailing from Amsterdam, Dirk Hartog made a simple navigational error and instead of reaching Java, he discovered the coast of what is now the state of Western Australia. But the Dutch East India Company was very secretive about such discoveries and it would be decades before maps with these details were released.

The most significant of these voyages, seen from a geographical perspective, were those of Abel Tasman. In 1642, Tasman sailed from Mauritius along a line south of Australia that led him to an island he named Van Diemen's Land (today's Tasmania). Continuing east, he became the first European to reach what is now New Zealand, before discovering Tonga and Fiji. In early 1644, he sailed again, this time turning south from New Guinea, missing Torres Strait (as had Janszoon) and entering the Gulf of Carpentaria. Heading west, he charted hundreds of miles of previously unknown coastline, naming it New Holland. In the aftermath of Tasman's voyages, maps began to appear, but they only showed limited details. It was not until 1663 that the French scientist Melchisédech Thévenot produced a map that revealed the range of Dutch discoveries, including much of New Holland.

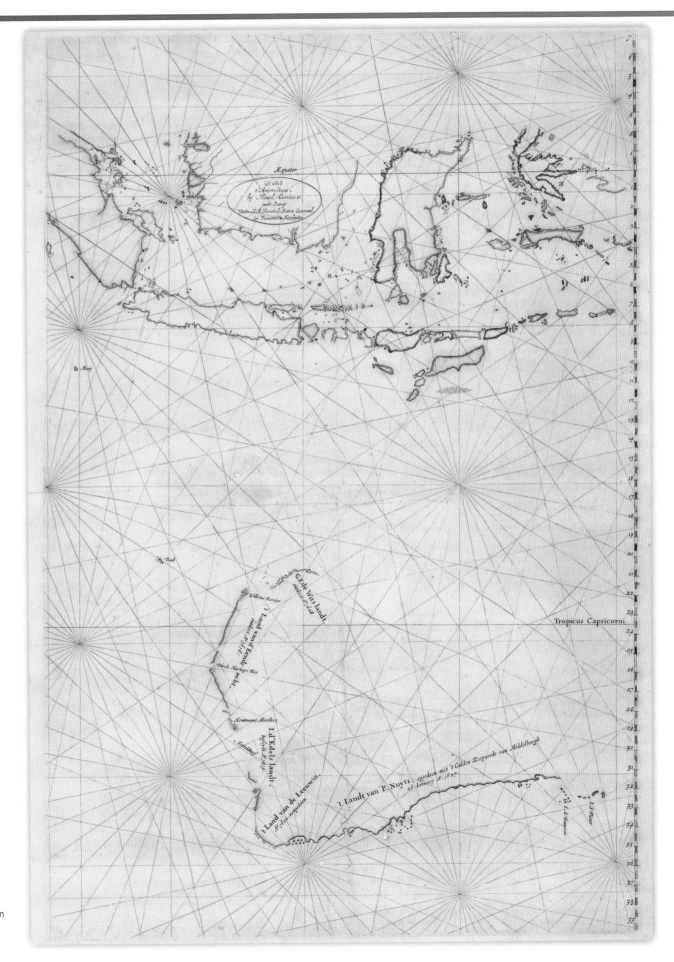

Left: A map of Abel Tasman's
voyages of 1642 and 1644,
undertaken in the service of
the Dutch East India Company.
Note that Australia extends down
to include Van Diemen's Land,
today's Tasmania.

The Dutch East India Company continued to dominate the exploration of the region for the next century, although the English buccaneer William Dampier visited New Holland twice and produced numerous Pacific maps – including one that depicted the trade winds.

The around-the-world voyages of Britons Samuel Wallis and Philip Carteret, which began in 1766, launched a new age of Pacific discovery. Becoming separated by bad weather near the Strait of Magellan, the two men crossed the ocean independently, with Wallis becoming the first European to land in Tahiti, and Carteret, who took a more southerly route, discovering Pitcairn Island. Three months after they had departed from England, an expedition under Louis-Antoine de Bougainville sailed from France, on its way around the world discovering the Great Barrier Reef. The charts produced on all three of these voyages would quickly become outdated, because before the three vessels had all successfully returned, the greatest navigator of the age had also headed towards the South Pacific.

James Cook commanded three expeditions between 1768 and 1779, and he made more discoveries and compiled more charts than any previous explorer of the Pacific. Already an experienced surveyor, on his first voyage he charted the coasts of New Zealand, made the first sighting of eastern Australia and confirmed the existence of a strait between that continent and New Guinea. Leaving England again the year after his return, he made the first crossing of the

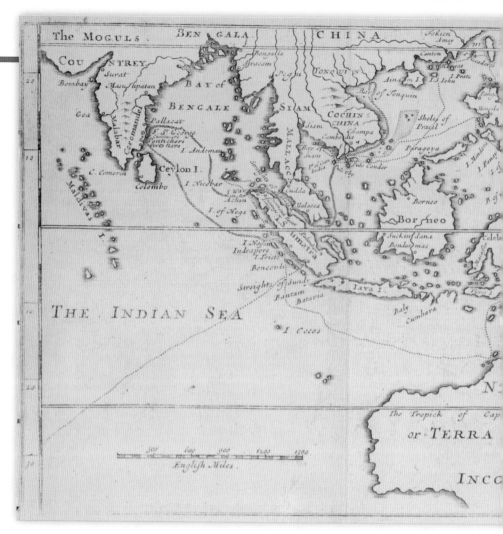

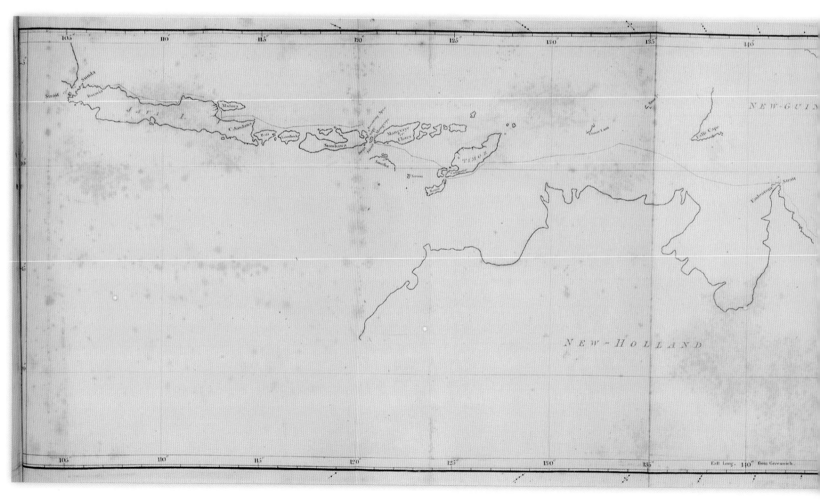

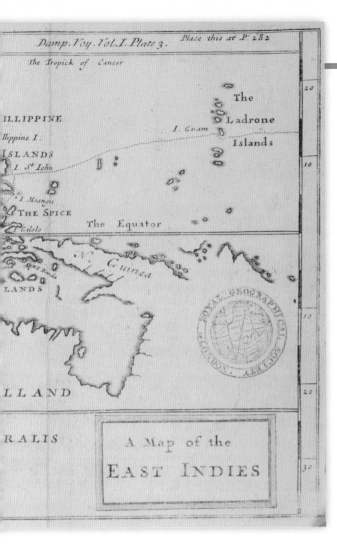

Antarctic Circle, established a record farthest south of 71°10′ S, and discovered places he called New Caledonia, Norfolk Island and the South Sandwich Islands. On his third voyage, Cook located and charted Iles Kerguelen, discovered Raratonga, the Cook Islands and the Hawaiian Islands (which he named the Sandwich Islands), and then passed through the Bering Strait before returning to Hawaii, where he was killed.

Cook's charts were the highest quality maritime maps yet produced, not only because of his own masterful skills, but also due to the recent development of the sextant and Harrison's chronometer. The charts were so complete that his successors had little to add regarding the locations he mapped until entirely new techniques were developed.

Six years after Cook's death, the French navigator Jean-Françoise de Galoup, Comte de la Pérouse, made landings at sites throughout the Pacific, from Easter Island to Kamchatka. In early 1788, having sent many of his charts back to France, he sailed from Botany Bay but was never seen again. It was only decades later that it was discovered that his ship had been wrecked off the island of Vanikoro. La Pérouse's extensive itinerary and output meant that his charts offered invaluable additions to those produced during Cook's expeditions.

Perhaps the most remarkable nautical drawings of all, considering the conditions under which they were produced, were those of William Bligh, captain of the British ship HMAV *Bounty* in 1789. Following the infamous mutiny, Bligh and 18 loyal seamen were set adrift in the ship's launch. During the next 47 days, Bligh navigated approximately 3,600 nautical miles (6,660 km) to Timor, with only one stop. Throughout the journey, which is considered one of the most remarkable accomplishments in the history of open-boat travel, Bligh kept a detailed log and made sketches of his course.

Left: The buccaneer William Dampier's map of the East Indies and New Holland from his 1697 book *A New Voyage Round the World*. Note that New Guinea is attached to New Holland.

Below: William Bligh's chart of his amazing open-boat journey in the launch of *Bounty* in 1789. The voyage began near the island of Tofua and ended at Timor in the Dutch East Indies.

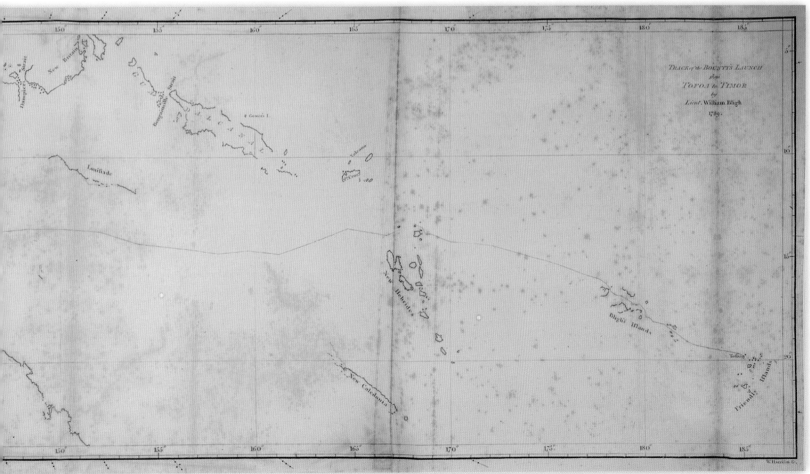

# Colonial Rivalry
## and Mapping in
# North America

Above: Every five miles along the Mason-Dixon Line was a "crownstone" boundary marker with the coats of arms of the Penns and Calverts on opposite sides. This marker is located today in Marydel, Maryland.

Left: Guillaume Delisle's Carte du Canada, which emphasized the vastness of New France at the expense of the British colonies. This version was published in Amsterdam in 1708.

given to the newly incorporated Hudson's Bay Company in 1670). French maps emphasized that country's control of the continent's heartland, known as Louisiana (La Louisiane to the French). Guillaume Delisle's "Carte du Canada" of 1703 is intriguing in its use of comparative sizing, because the British colonies are minute in contrast to New France, while the area controlled by the Hudson's Bay Company is a wasteland.

One popular cartographic term for North American maps was the Latin form "America Septentrionalis", and numerous maps – with different emphases – shared it. One of the first was produced in 1636 by Henricus Hondius (son of Jodocus), which showed California as an island. In the 1650s, Nicolas Sanson produced a French version, L'Amerique Septentrionale, which was modified and re-engraved many times.

Perhaps the best-known map of this name, by Henry Popple, appeared in 1733. The massive original measured nearly 2.4 × 2.3 m (7 ft 10 in × 7 ft 6 in), and when produced as an atlas it consisted of a hand-coloured index map and 20 uncoloured regional maps. In reality, the map was an exercise in political propaganda: Popple had been hired by Britain's Board of Trade to provide cartographic support to British territorial claims being disputed by the French and Spanish.

By the mid-1750s, Britain and her allies were headed yet again towards war with France and her allies, a conflict that would rage throughout Europe, North America, the Caribbean, Africa, India and on the high seas worldwide. That year, representatives from Britain and France met to try

I N THE SEVENTEENTH CENTURY, the longstanding European rivalry between Britain and France found a fresh outlet when both countries began to establish colonies and commercial companies throughout the world. Many of these new territories were in close proximity and a struggle for regional control soon dominated Anglo-French relations in India, the Caribbean and North America.

British and French maps of the time reflected these geopolitics, as the cartographers from both expanding empires attempted to illustrate their own domination. For example, British maps of North America often showed New France as a relatively thin strip between the 13 contiguous British colonies and Rupert's Land (the name given to the vast region that essentially consisted of the drainage basin of Hudson Bay, which had been

A MAP
of the BRITISH EMPIRE in
AMERICA
with the FRENCH and SPANISH
SETTLEMENTS adjacent thereto
by Hen: Popple.

to resolve the territorial conflicts in North America. Driven strictly by political considerations, the British dismissed Popple's map – the one they had commissioned – as highly inaccurate, arguing that it failed to show the true extent of British control.

However, in the same year, two other maps appeared that presented the boundaries more to the British government's liking. The first was by William Herbert and

Robert Sayer, who would later make major advances in mariners' charts. The title of their map clearly explained the British stance: "A New and Accurate Map of the English Empire in North America: Representing their Rightful Claim as confirm'd by Charters, and the formal Surrender of their Indian Friends; Likewise the Encroachments of the French, with the several Forts they have unjustly erected therein". In addition to the exaggerated size of the British

Above: Henry Popple's 1733 map America Septentrionalis. Despite its great size allowing for incredible detail, numerous aspects of the map do not conform to the correct size or shape, including the Great Lakes, the islands of the Caribbean and Florida.

colonies vis-à-vis New France (in the manner of Delisle's map, but from the other side's point of view), Virginia, the Carolinas and Georgia extended far to the west of the Mississippi river, which was territory that had been explored or previously claimed not by Britain but by France or Spain.

The second map has been described as the most important in the history of North America. Remarkably, it was produced by a medical practitioner, John Mitchell, who had no previous experience of mapping. A native Virginian, Mitchell had moved to England for health reasons in 1746, when England was again at war with France (the War of the Austrian Succession, 1740–48). Mitchell disliked the French because some of his possessions had been taken by a French privateer, and he decided to produce a map so that the British public could see the extent of the threat they posed to Britain's colonies.

Working from publicly available sources of information, Mitchell completed his first draft in 1750. At that point, the Board of Trade, impressed by what he had accomplished, retained Mitchell to make a new map, making use of official reports and charts. Further, the Board requested that the colonial governors send to Mitchell detailed maps and boundary information, particularly showing French "incursions".

By 1755, Mitchell had completed his map, which measured nearly 2 × 1.4 m (6 ft 6 in × 4 ft 5 in). Although not as large as Popple's map, it showed the Atlantic colonies and Canada in significantly greater detail, and numerous annotations provided a great deal of additional information. Mitchell always gave the benefit of the doubt to British claims, such as the boundaries dividing Massachusetts from Quebec, and, like Herbert and Sayer, he extended the western boundaries of the British colonies all the way to the Pacific Ocean. Mitchell also indulged in the Delisle-like selective expansion and contraction of geographic space.

By the time a second edition of Mitchell's map appeared in 1757, Britain was yet again at war with France. It was only with the publication of the fourth edition, in 1775, that the British victory at Quebec was included on the map, coincidentally within months of the beginnings of the American War of Independence (1775–83).

Right: New France, or what is today Canada, from *Carte de l'Amerique Septentrionale*, produced by Jean-Baptiste-Louis Franquelin in the late seventeenth century. He also served as royal hydrographer.

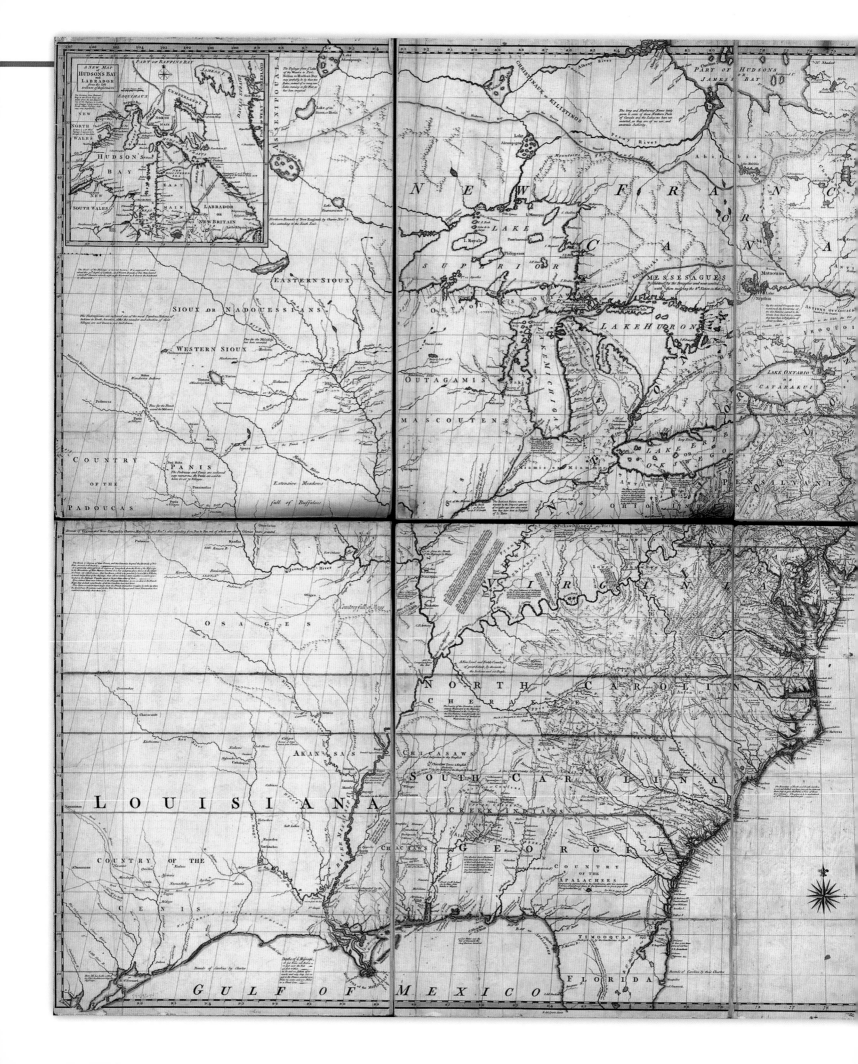

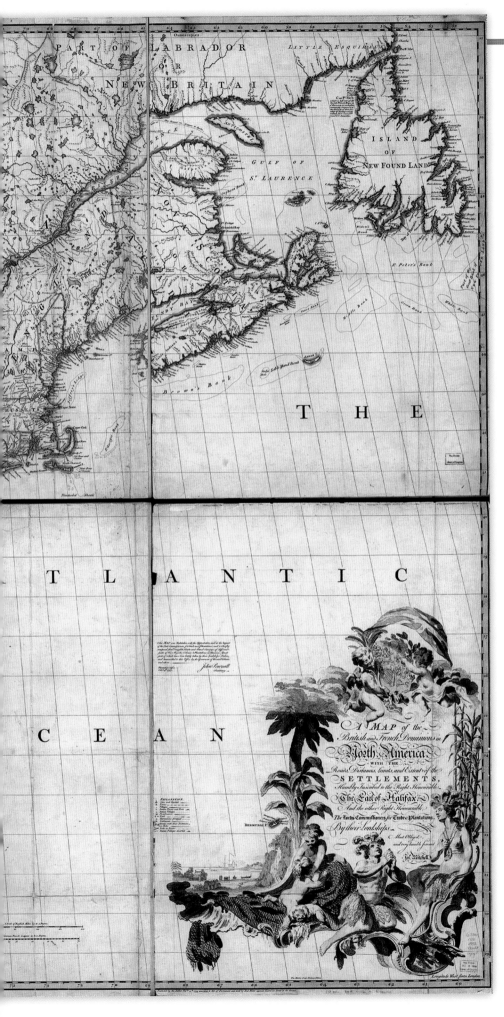

Ironically, eight years later, when the British and Americans negotiated a peace treaty, copies of Mitchell's map – designed to show Britain's control of North America – were used to determine the new country's final boundaries, at Britain's expense. For example, on his copy, the British diplomat Richard Oswald added lines in red ink to show the British interpretation of the new border. That map, later given to King George III (reigned 1760–1820), came to be known as the "red line" map (a name also occasionally applied to other copies that were annotated in red at the time or shortly thereafter). As the agreed record of the international boundary, the map has been consulted for disputes between Canada and the United States as recently as the 1980s.

Throughout the Seven Years' War (1756–63, often known as the French and Indian War in the US, although that conflict began in the region a couple of years earlier) a series of maps and charts recorded different aspects of the conflict. One book, *A Set of Plans and Forts in America* by Mary Ann Rocque, contained diagrams of 30 different forts – some British and some French – and served as a visual celebration of the British victory that expelled the French from North America.

A different, decades-long boundary conflict was resolved with less bloodshed. The dispute, between the powerful families of Pennsylvania and Maryland (the Penns and the Calverts), owed its existence to the fact that the charters for those two colonies, as well as that for Delaware, were in disagreement as to what the exact borders were. Maryland even claimed territory that included Philadelphia, Pennsylvania's main city. After violence erupted between settlers in the 1730s, the British crown finally imposed a resolution on the colonies in 1760. As a result, the families commissioned the English team of astronomer Charles Mason and surveyor Jeremiah Dixon to survey and establish the precise boundaries of the affected colonies. Between 1763 and 1767, the pair surveyed the line between Maryland and Delaware, and then 393 km (244 miles) of the line between Pennsylvania and Maryland before they were forced to quit by opposition from local Native Americans.

In the 1770s the project was restarted and by 1784 the survey was mostly complete, having been extended west to include much of the line between Pennsylvania and Virginia. The outcome was what became known as the Mason–Dixon Line. After the Missouri Compromise (about the regulation of slavery) in 1820, the line came to symbolize the cultural difference – including the status of slavery – between the northern and southern states.

Left: The copy of John Mitchell's "red line" map held at the Library of Congress. This version was the first edition of 1755. Note that the sizes and shapes of lakes and coastlines are more accurate than those on Popple's earlier map.

For nearly a century, the Hudson's Bay Company had controlled Rupert's Land to the northwest of New France. Although it was important for the Company's traders and agents to be familiar with the geography of their massive domain, particularly its water routes, such information was generally kept secret (in the manner of the Dutch East India Company). That way, any traders from New France would gain no advantage when competing for the same pelts, fish and other resources. Although France had been expelled from eastern North America after losing the Seven Years' War, the Hudson's Bay Company remained in competition with long-established French-speaking traders, as well as from a rival, Montreal-based group of fur traders known as the North West Company.

One of the Hudson's Bay Company's hopes of further profit was to find valuable metals, and in 1770 Samuel Hearne was sent from the main port at Churchill to seek copper deposits and to find a low-latitude sea route to the north and west, a hypothetical waterway the English had long called the Northwest Passage. The next year, after passing through a series of lakes and rivers, he descended the Coppermine river to the Arctic Ocean, becoming the first European to stand on the Arctic coast of the North American continent.

Throughout this journey, Hearne kept "a large skin of parchment that contained twelve degrees of latitude North and thirty degrees of Longitude West of Churchill Factory". On this parchment he "sketched all the West Coast of the Bay ... but left the interior parts blank, to be filled up during my journey". He also made extensive notes about lakes, rivers and other physical features. In 1772, Hearne filed a report with the Hudson's Bay Company, along with the first map ever made of the interior of sub-Arctic Canada. At the same time, he produced a second map, and it was this that he included when his expedition account was published more than 20 years later, showing the public for the first time the geography of what was once described as that "rascally and ungrateful land".

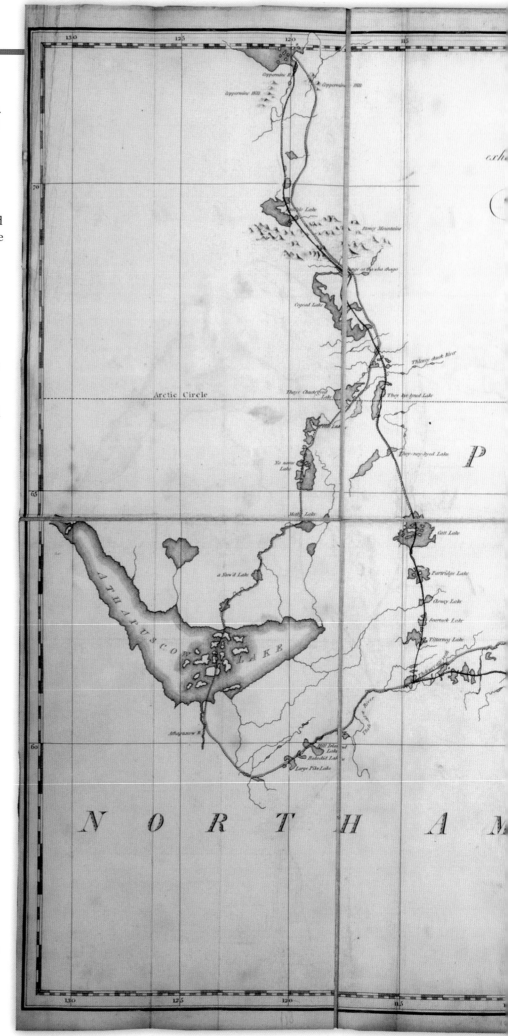

Right: The route map of Samuel Hearne's journey to the edge of the Arctic Ocean in 1770–72. This first appeared in his 1795 book *A Journey from Prince of Wales's Fort in Hudson's Bay to the Northern Ocean*.

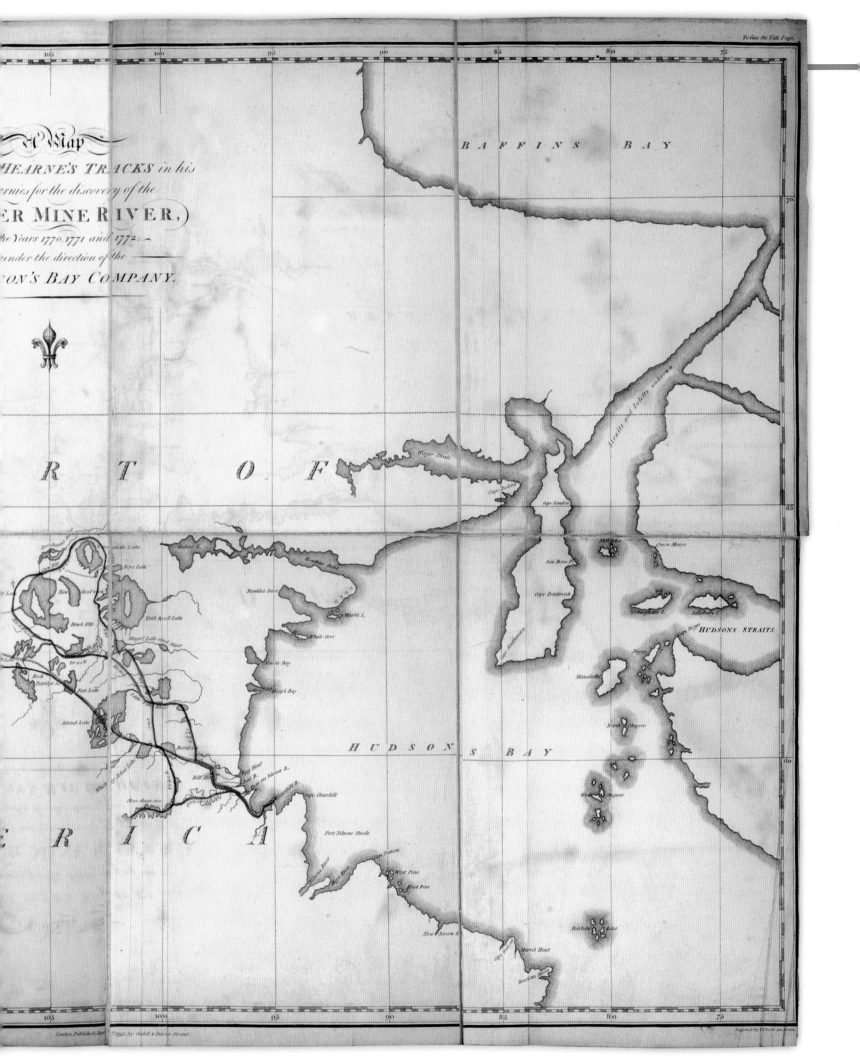

A Map
HEARNE'S TRACKS in his
Journies for the discovery of the
ER MINE RIVER,
the Years 1770, 1771 and 1772
under the direction of the
SON'S BAY COMPANY.

BAFFINS BAY

R T   O F

Wager Strait

Cape Dobbs

Cape Comfort

Mill Lake

Queen Marys

Sea Horse Pt.

Chesterfield Inlet

Cape Southampton

Cape Pembroke

Nottingham I.

Salisbury I.

Rankins Inlet

Cape Digges

HUDSONS STRAITS

Marble I.

Diggs I.

Whale Cove

Mansfield I.

Knaps Bay

Quoy's Bay

North Sleepers

HUDSONS BAY

Egg River

Seal R.

Cape ree hhewn R.

Churchill R.

Cape Churchill

West Sleepers

Stoe-than-nee

Port Nelsous Shoals

Nelsous River

Hay River

Cape Tatnam

West Pine

East Pine

Belchers Isles

New Severn R.

Cape Jones

Marvel Head

ERICA

# James Rennell:
## Mapping India, Africa and Ocean Currents

Above: Portrait of James Rennell produced in 1799. Rennell has been called "the father of oceanography", "the father of Indian geography" and "the father of modern cartography".

I**N THE HISTORY OF CARTOGRAPHY**, few individuals stand out for their work in so many geographical regions and aspects of science as James Rennell. Born in Devon in 1742, Rennell went to sea at the age of 14, learned maritime surveying and then, at the end of the Seven Years' War, received a commission in the Bengal Army as an engineer.

Britain's successes in the war had effectively secured for the British East India Company the right to administer Bengal and collect its land revenues. However, the company possessed little geographical information about India's interior because previous surveys had not made use of the most recent technology and instruments. With a pressing need for accurate topographic information to aid with troop movements, the transport of merchandise and the collection of taxes, a survey of Bengal was ordered, which began with the rivers.

Equipped with quadrant, compass and chain, Rennell began a thorough and scientific survey of the major river systems, roads, plains, jungles, mangrove forests and mountains. So impressive was his work that, in 1767, Robert Clive appointed him surveyor-general in Bengal, at the age of just 24. For the next decade, from his headquarters in Dacca, Rennell oversaw the management of the geographical information and the production of numerous maps.

Having never fully recovered from a severe wound received in an ambush, Rennell retired in 1777 and returned to London, where he produced his masterful *A Bengal Atlas*, which included an index map, 20 regional maps and a page of "views". Published in 1780 and based on more than 500 individual surveys, the atlas was extremely significant due to its large scale and the unprecedented accuracy of its details – even locations of violent incidents (such as his own wounding) were included, represented by crossed swords. Each map was produced using the same scale, with the district centred

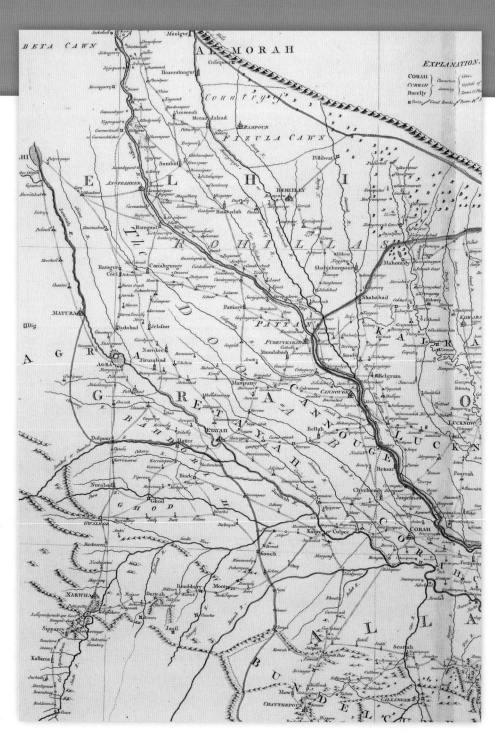

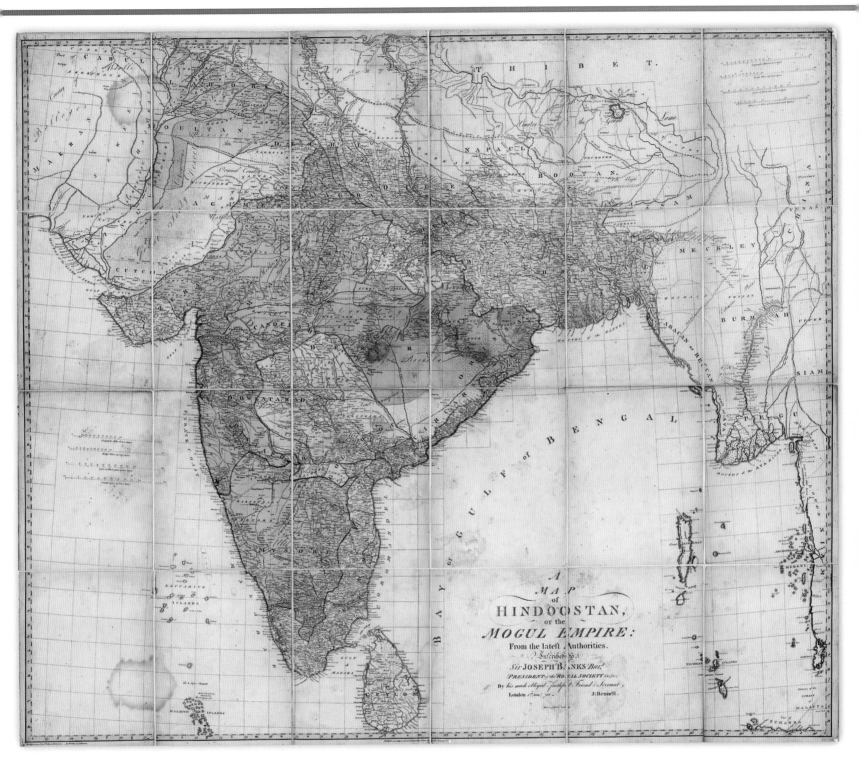

on the sheet, although the orientation of the sheets varied in order to accommodate the different shapes and sizes of the regions. Rennell's maps were such a major advancement over their predecessors that they remained the definitive administrative maps of Bengal for almost five decades.

Yet, what is arguably Rennell's greatest map was still to come. Published twice with multiple editions, the Map of Hindoostan was considered the first accurate chart ever made of the whole of India. Issued initially in 1782 on two large sheets totalling about 1.5 × 1.2 m (58 × 48 in), in 1788

the map increased to four sheets with a scale of 3.8 cm (1.5 in) per equatorial degree. Intriguingly, Rennell divided these maps according to the Mughal provinces defined by the sixteenth-century Mughal emperor Akbar the Great (reigned 1556–1605), establishing a conceptual equivalency between British India and the Mughal Empire. Accompanying the map was a book entitled *Memoir of a Map of Hindoostan; or the Mogul Empire*, which described in minute detail all the sources used for the maps and the criteria by which they had been accepted or rejected.

Above: Rennell's masterpiece – A Map of Hindoostan, or the Mogul Empire. This is the second edition, produced in 1788 and issued on four sheets.

Opposite: Part of one of the regional maps of *A Bengal Atlas*, showing the area of Delhi and Agra. Rennell's atlas was unusual in that it combined exceptional detail with remarkable clarity.

Rennell was also greatly interested in the exploration of Africa. Using diaries, surveys and reports by several noted explorers, he produced maps for their published travel accounts. These included Mungo Park's route up the Gambia river to the Niger river; an annotated map of North Africa, for the account of Friedrich Hornemann's travels before his mysterious disappearance; and James Kingston Tuckey's route on his disastrous expedition up the Congo river. Rennell's commentary accompanying the maps also demonstrated the way he applied logical analysis to reach his conclusions, even when, as in the case of the Niger, they were eventually proved to be inaccurate.

In 1777, on the voyage back to England from India, Rennell's journey included an extended stay in southern Africa, where he became interested in the unusual pattern of ocean currents there. Using a variety of instruments in a small boat, he attempted to chart the depth, direction and speed of the currents, and the next year he published a memoir and chart about what is today known as the Agulhas Current. This pioneering contribution to the science of oceanography initiated serious studies in hydrography that he would continue for the rest of his life. In 1793, Rennell was the first to offer an explanation for the

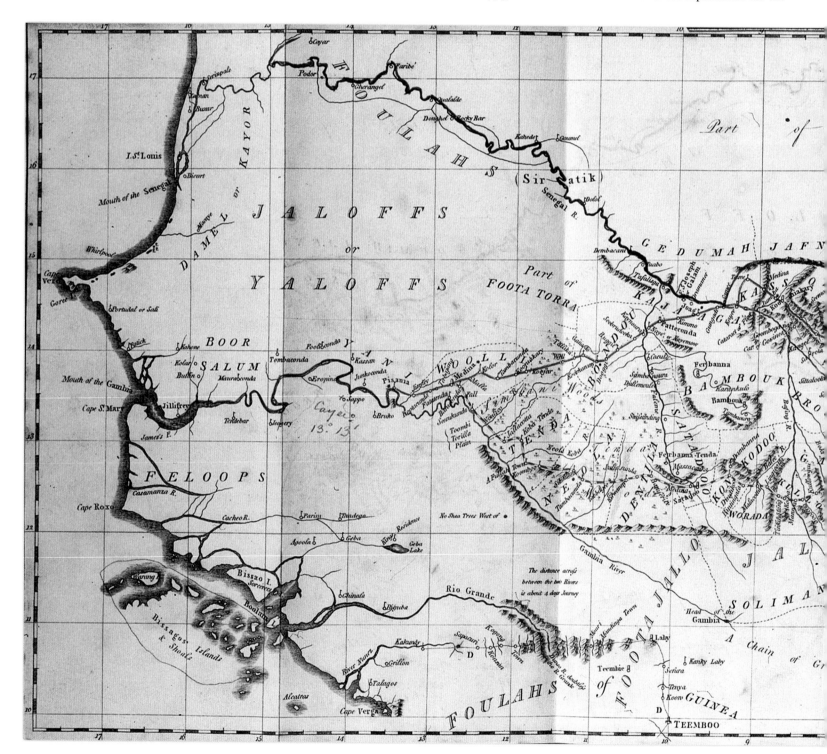

causes of the dangerous occasional northern current found to the west of the Scilly Isles. He later spent many years gathering information from the logs and notebooks of his naval acquaintances (including Alexander Dalrymple and Francis Beaufort, both hydrographers of the Admiralty, and the explorers William Bligh and Matthew Flinders), which he sifted and assimilated in order to chart the currents of the Atlantic Ocean.

Rennell died in 1830 and was buried in Westminster Abbey. His *An Investigation of Currents of the Atlantic Ocean*, which included

numerous charts and is often considered to form the historical basis of the study of currents, was published posthumously in 1832. The work he produced on the Gulf Stream is judged not to have been surpassed until more than a century after his death.

Below: Rennell's map of the route taken by Mungo Park in his first expedition up the Gambia river to the Niger river. This accompanied Park's expedition account *Travels in the Interior Districts of Africa*.

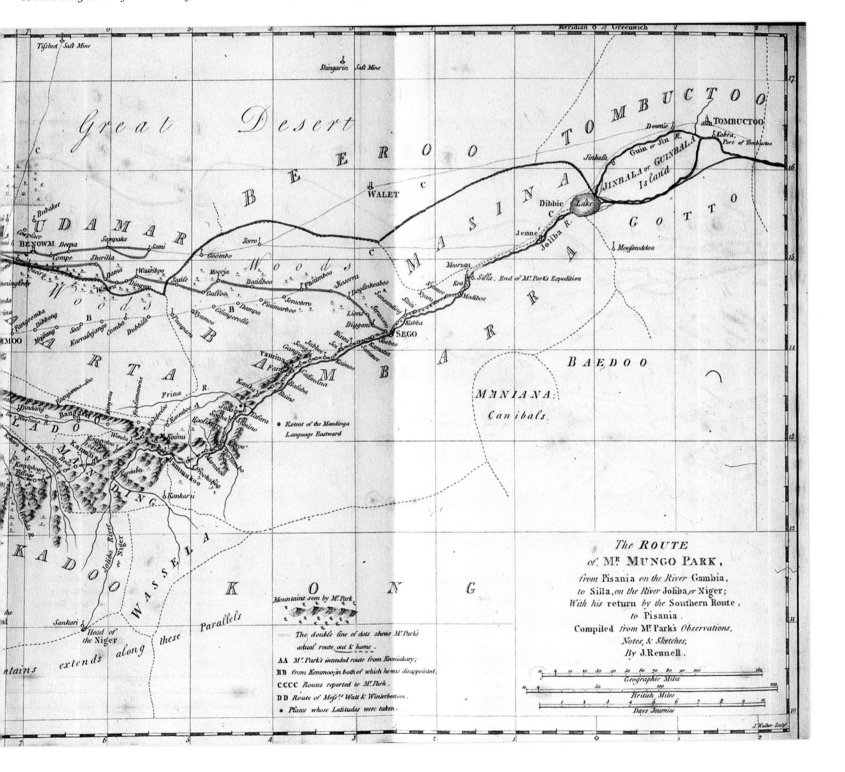

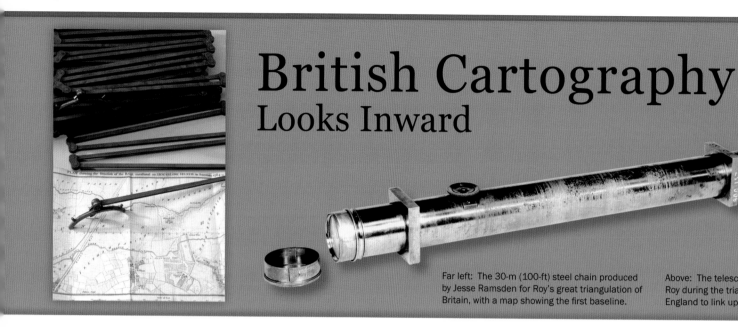

# British Cartography
## Looks Inward

Far left: The 30-m (100-ft) steel chain produced by Jesse Ramsden for Roy's great triangulation of Britain, with a map showing the first baseline.

Above: The telescope used by William Roy during the triangulation of southeast England to link up with that of France.

THROUGHOUT THE EIGHTEENTH CENTURY, committees or commissioners of the British government, the Royal Navy and the British East India Company were each mightily concerned with the surveying of lands and waterways around the world, and the production of accurate maps and charts for them. By contrast, there was little similar official interest about Britain itself – the topographical maps of British counties, roads or towns tended to be privately produced by those whose first interests were commercial.

The government became involved in British cartography in the aftermath of the Jacobite Rising of 1745–46. Following the government's victory at Culloden, David Watson, the deputy quartermaster-general, began a military survey of Scotland as part of a strategy to suppress any future uprisings. This included building forts connected by new roads and mapping the entire country, so that troops could move swiftly and accurately.

Watson's primary assistant was a 21-year-old Scots surveyor and cartographer, William Roy. During the next five years, Roy and his colleagues conducted traverses throughout the Highlands using chains to determine distance and circumferentors (surveyor's compasses) to measure horizontal angles. When the survey was finished in 1752 – at a scale of 2.5 cm (1 in) to 914 m (3,000 ft) – it was extended to southern Scotland, until the process was interrupted by the Seven Years' War.

At the beginning of 1756, Roy – now commissioned in the army – started work on a reconnaissance survey in Kent and Sussex, mapping roads and landmarks in case of a French invasion. After the Seven Years' War, he was appointed to the post of surveyor-general of the coasts and engineer for directing military surveys in Great Britain, but his requests in 1763 and 1766 for official surveys of all of Britain were turned down. Instead, such mapping remained a private, commercial venture.

Roy's chance finally came two decades later, when Jacques-Dominic Cassini proposed that a triangulation of southeast England be linked to the French one to determine the difference in longitude between Greenwich and Paris. Roy promptly ordered the finest possible instruments – including a 30-m (100-ft) steel chain and a theodolite with a base circle of 91 cm (3 ft) – from Jesse Ramsden, the foremost instrument-maker of the period. Unfortunately, the theodolite took three years to produce, and after establishing a baseline of 8 km (5 miles), Roy simply had to wait. When the theodolite was eventually delivered in 1787, the triangulation was carried out from Hounslow Heath to the coast of Kent and connected to the French observations. In a test of verification, an actual measurement was found to differ by only 71 cm (28 in) from what the triangulations had indicated the distance would be.

In 1790, Roy died, just a year after the triangulation

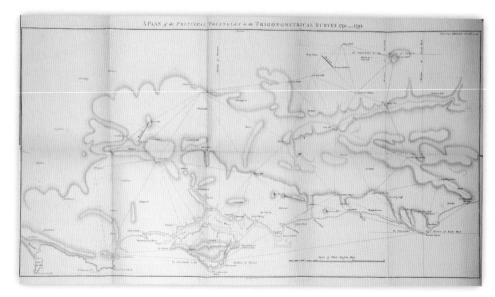

Left: A diagram showing the principal triangles in the south of England during the first years of the Trigonometrical Survey. The triangles had reached Land's End by 1795.

Opposite: The map of Lancashire from John Cary's *New and Correct English Atlas* of 1787. The data obtained by the Ordnance Survey and used by private cartographers made possible even more thorough maps in the following years.

was completed. However, his work had laid the foundation for a full national survey based on firm geodetic principles, and it did not take long to fulfil this legacy. In 1791, the Trigonometrical Survey was established. Using the Ramsden theodolite, and adhering to Roy's systems and standards, the new organization began expanding Roy's work. By 1795, a chain of triangles had been extended from London to Land's End. In 1801, the Trigonometrical Survey was officially renamed the Ordnance Survey. That same year, its first county maps – at a scale of 2.5 cm (1 in) to 1.6 km (1 mile) – were released, with Kent published privately by William Faden, and Essex shortly thereafter the first actually published by the Ordance Survey. What became known as the Principal Triangulation of Great Britain was finally completed in the 1850s.

The establishment of the Ordnance Survey did not immediately put an end to maps privately produced in Britain. In fact, many improved greatly because of their use of Ordnance Survey data. Moreover, there remained an appetite for maps with many different purposes; for example, the cartographer John Cary produced county maps, road maps, itineraries, canal plans, geological maps, astronomical charts and globes. His *New and Correct English Atlas* (1787), a set of county maps, exemplified cartographic accuracy and exceptional engraving, and it was reissued 21 times. In 1794 he was named surveyor of roads for the General Post Office. This allowed him to collect even more information and, when combined with material from the Ordnance Survey, enabled him to make *Cary's New Itinerary* of 1798 one of the most thorough maps of its era.

By 1795, at a point when the Trigonometrical Survey's triangulation had addressed some concerns about a French invasion, it had also become apparent that there was a need for reliable sea charts for the same defence matters. That year Alexander Dalrymple was appointed the first Hydrographer of the Admiralty. His responsibilities included bringing order to the many, varied charts of coastal and oceanic waters and then compiling new ones. In 1811, the publication of *Charts of the English Channel* by Dalrymple's successor, Thomas Hurd, marked a step forward in that mission. The success of the atlas of 31 charts helped to launch what would become known as the Grand Survey of the British Isles – a magnificent series of charts, the production of which, under the fourth Hydrographer, Francis Beaufort, was in many ways comparable to the mapping of the Ordnance Survey.

Right: One of the first Ordnance Survey maps showing part of the county of Kent.

# CHAPTER 4

# Maps in the *Age of Empires* and *Nationalism* (1800–1914)

Right: The western part of the map of the previously unknown interior of central Africa produced from observations made by Henry Morton Stanley during his crossing of Africa in 1874–77. This appeared as both one map and two in different versions of Stanley's *Through the Dark Continent*. It included the first-ever tracing of the course of the River Congo and its relationship to Lake Tanganyika, as well as the first outline of Lake Victoria by a European who had actually followed its entire shoreline.

Map showing the
WESTERN HALF
OF
EQUATORIAL AFRICA
AND THE
EXPLORATIONS BY LAND AND WATER
OF
HENRY M. STANLEY
IN THE YEARS 1874-77.

English Miles

Geographical Miles

M? Stanley's Route

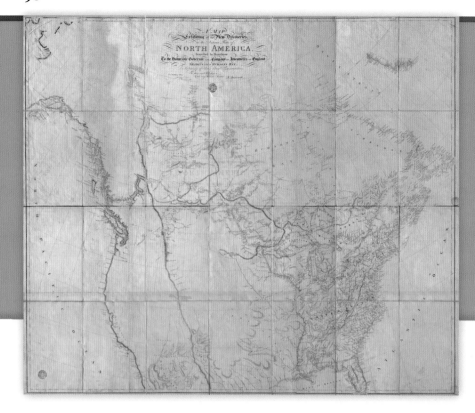

# The Opening of North America

Left: Aaron Arrowsmith's Map Exhibiting All the New Discoveries in the Interior Parts of North America. The 1802 version of this map was probably the most important map in the planning of Lewis and Clark's expedition.

Opposite: Alexander von Humboldt's Carte du Mexique – published in 1811 based on Humboldt's observations and information gathered from libraries and archives – was a seminal map of not only what is today Mexico but also the American southwest, Texas and the Rocky Mountains.

IT WAS NOT LONG AFTER THE END OF THE AMERICAN WAR OF INDEPENDENCE (1775–83) that the number of pioneers passing over the Appalachian Mountains into the lands to the west changed from a trickle to a flood. Some of these territories had already seen settlements spring up – such as Boonesborough, in the part of Virginia that would become Kentucky – but the movement of people had been slowed after the Seven Years' War (1756–63), when a royal proclamation set aside for the native peoples much of the area that had previously been claimed by France.

In 1783 the Treaty of Paris recognized the United States (US) as including the territory that stretched from Canada in the north to Florida in the south, and from the Atlantic Ocean in the east as far as the Mississippi river in the west. Eventually, the new states with claims in the west of the region ceded those lands to the US government, which, in turn, created territories (such as the Northwest Territory) that were opened for expansion and settlement.

The concept of what was the "West" changed radically in the American mind in 1803. Napoleon's military and diplomatic successes in Europe had caused much of the area west of the Mississippi to pass from Spanish to French control, but the costs of his wars made the control of a North American empire financially unfeasible, so the French offered to sell it. The subsequent Louisiana Purchase cost $15 million, transferred 2,147,000 sq km (828,800 sq miles) from France, doubled the size of the US and opened vast new vistas for American exploration, settlement and commerce – as well as mapping.

Right: A map of the regions traversed by Lewis and Clark, with annotations in brown ink by Meriwether Lewis, showing the northern section of the Mississippi river and rough outlines of Lake Michigan and Lake Superior.

Of course, the US was not alone in wanting to investigate and record new lands. More than a decade earlier, in 1789, Alexander Mackenzie of the North West Company set out to cross Rupert's Land. But when he followed what is now named the Mackenzie river, he reached not the Pacific Ocean but the Arctic Ocean, becoming only the second European to visit that mysterious coast. Three years later Mackenzie tried again, and after canoeing along rivers and trekking over mountains, he emerged on the Pacific – the first European to have crossed the continent north of Mexico. However, it was not until 1801 that Mackenzie's map of his journeys was finally published.

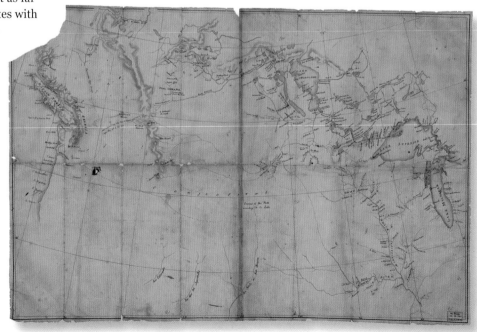

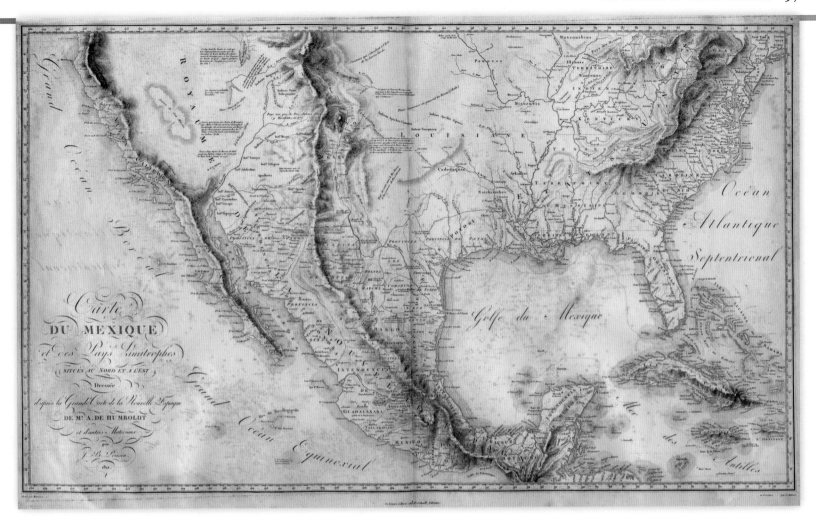

While Mackenzie was in the field, so was George Vancouver, a Royal Navy surveyor who had twice sailed with Captain Cook. In 1791–95 Vancouver led a series of coastal and river surveys from Alaska to California, producing highly accurate charts of the bewildering array of islands, inlets and rivers along the coastline – many of these charts remained the standard well into the twentieth century.

Mackenzie's and Vancouver's maps were both used in 1802 by the renowned London cartographer Aaron Arrowsmith to update his work, A Map Exhibiting All the New Discoveries in the Interior Parts of North America. This was the most accurate cartographic representation of the American West then available, and it was a key source for President Thomas Jefferson, who had plans to send out a transcontinental scientific and exploring expedition.

Jefferson's primary goal in sending out an expedition was to find a water passage across the continent, not only for the purposes of commerce but also in support of territorial expansion. In addition, he wanted the expedition members to make scientific observations and collect geological, botanical and biological samples; to investigate the language, laws and customs of the Indian nations; and to map the topography all the way to the Pacific.

In 1803 Jefferson entrusted the leadership of the endeavour to his personal secretary, Meriwether Lewis, who selected William Clark as his co-commander. In 1804–06, the Corps of Discovery, as the party was known, travelled by boat, horse and foot from St Louis to the Pacific and back, returning with an unprecedented wealth of new knowledge. Clark drafted all but three of the 140 maps drawn during the expedition. The main "track map" measured approximately 0.9 × 1.5 m (3 × 5 ft) and it was not until a decade later that a reduced-size version finally appeared of what was the most accurate cartographic report yet of the region.

The Lewis and Clark expedition led to many more ventures that essentially completed the map of the American West in the era prior to the American Civil War (1861–65). While the Corps of Discovery was still in the field, Lieutenant Zebulon Pike of the US Army ascended the Mississippi river into Minnesota in an unsuccessful attempt to find its source. The next year, he headed into territory that belonged to Spain, where, after making an unsuccessful attempt to climb a mountain in Colorado (today known as Pikes Peak), he and his companions were captured by the Spanish. Suspected of spying, Pike and his group were taken to Mexico and their notes and charts were confiscated before they were

Above: A portrait of Zebulon Pike (1779–1813), the American soldier and explorer who mapped much of the southern part of the Louisiana Purchase. Promoted to brigadier general in 1813, he was killed that year in military action at York (now Toronto).

Below: William Clark's map of the journey of exploration and science by the Corps of Discovery – commanded by Clark and Meriwether Lewis – on its historic crossing of North America from the Mississippi River to the Pacific Ocean and back.

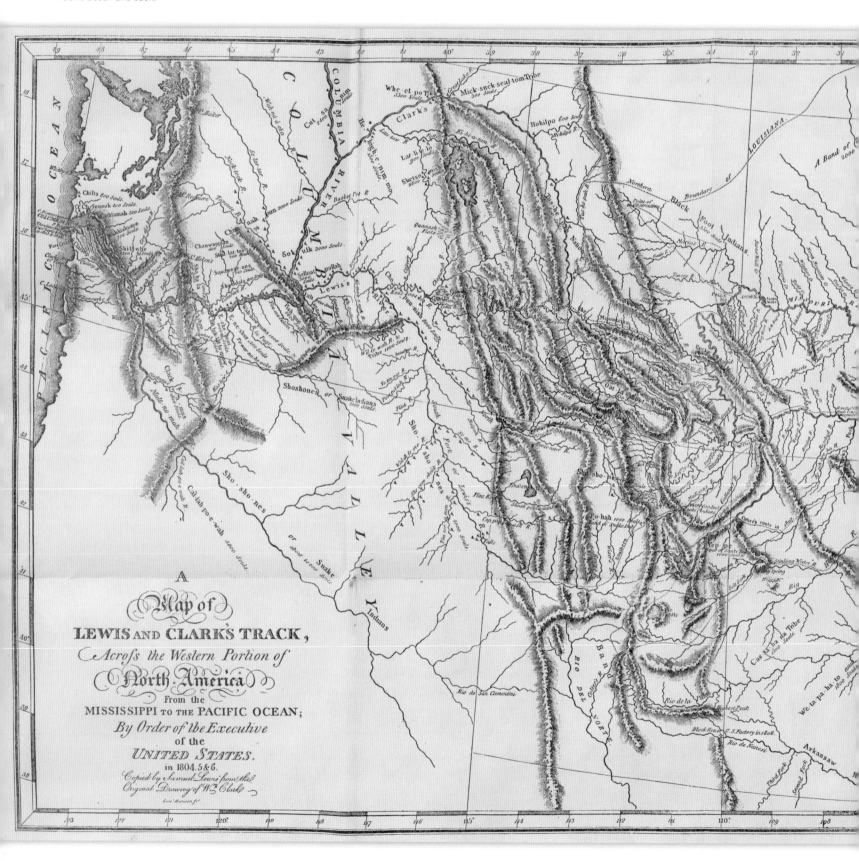

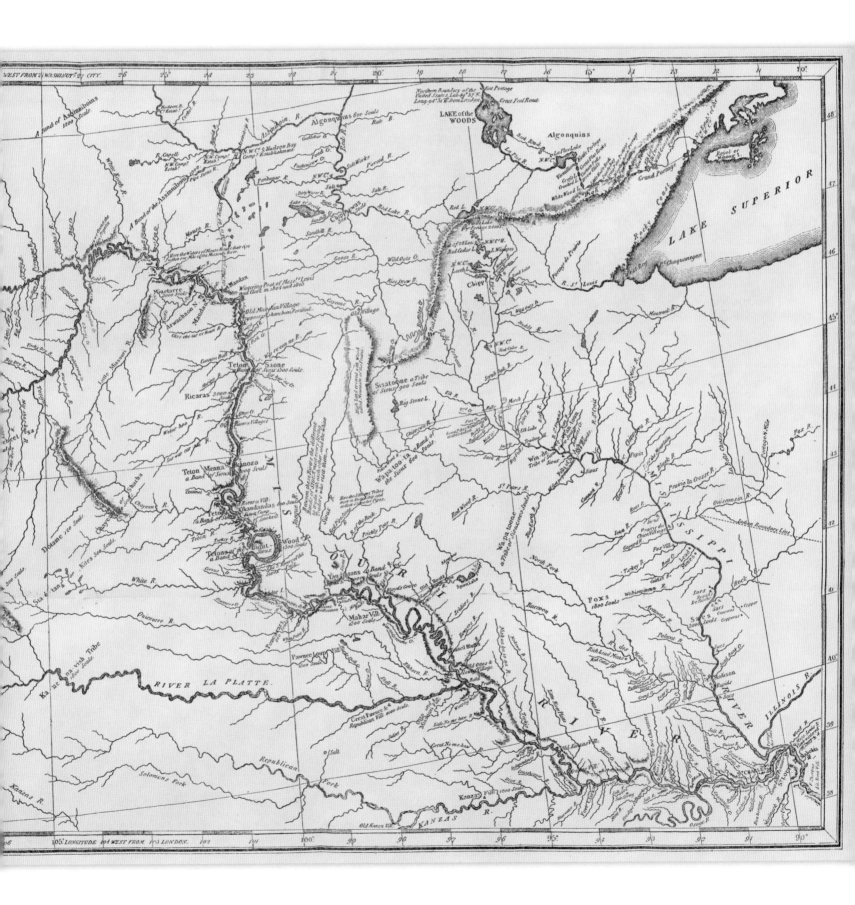

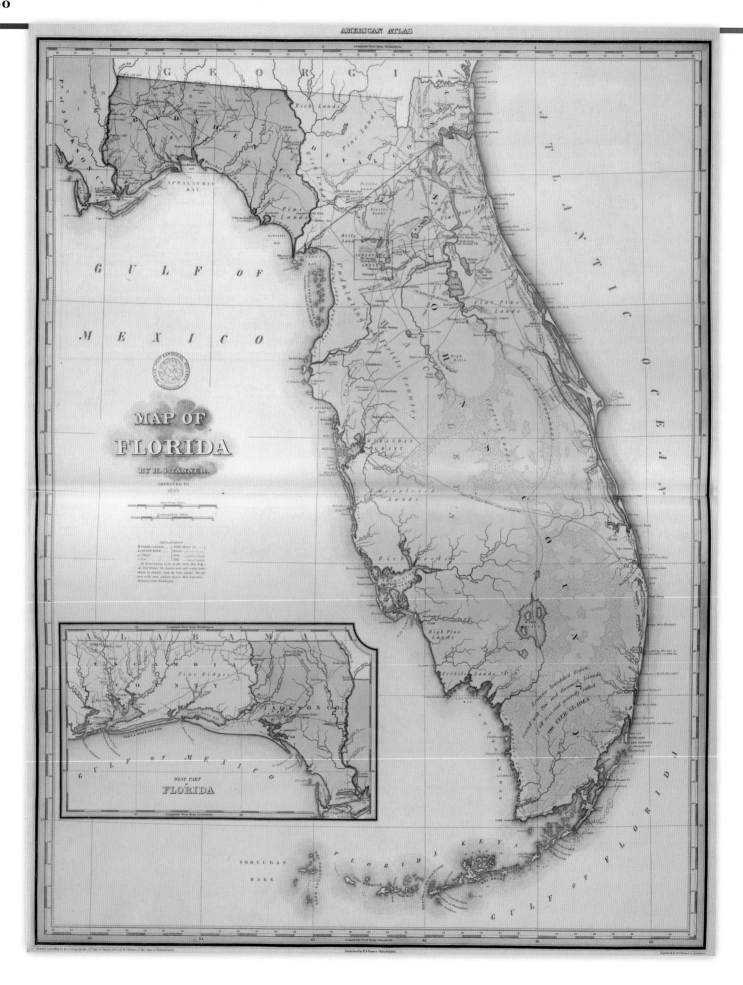

MAP OF
FLORIDA
BY H. S. TANNER.
IMPROVED TO
1825

eventually released. However, Pike was able to write an account from memory, published in 1810, which included maps of what he called Louisiana and New Spain.

Pike's maps used material from other sources. One of the likeliest was the famed Prussian scientist Alexander von Humboldt, who had visited Mexico in 1803 and produced Carte du Mexique, which introduced new topographical information to the American public about its neighbour. But even before Humboldt's map was officially released, Pike may well have used the original charts in preparation for his own expedition.

Following in the scientific tradition of Lewis and Clark, in 1819 the US Army sent an expedition westwards under Major Stephen Long. For the next year and a half, Long explored a series of rivers – the Missouri, Platte, Red and Arkansas. However, Long failed to find their sources, and he announced that the lands through which he travelled were unfit for cultivation. Although his maps corrected previous errors, he made a significant blunder by labelling one region the "Great American Desert". Known today as the Great Plains, this region includes some of the nation's most productive farming areas.

Perhaps the finest American cartographic effort of the nineteenth century produced by someone who was a map-maker but not an explorer was Henry Tanner's *A New American Atlas* of 1823. Built upon a combination of other men's original surveys and a large number of recent maps, the book's lavish production generally showed two or more states per hand-coloured, double-page map sheet, with all the maps of a uniform size and scale.

In 1820, while Long experienced hardships in the West, Moses Austin, an entrepreneur who had lost his fortune, travelled south with an audacious business plan. Eventually, he succeeded in convincing the Mexican authorities to allow him to establish an American colony in Texas. Although he died before he could launch his venture, his son, Stephen F. Austin, continued what Moses had begun. Believing strongly that a good map would encourage settlement, Austin spent six years gathering information and preparing preliminary copies; he sent these to Philadelphia, where the renowned map-printer Henry Tanner published the map to great acclaim in 1829. Seven years later, Tanner was preparing to reissue it when Austin wrote to him to request that "April 21, 1836" be imprinted on the map to commemorate the Battle of San Jacinto, at which Sam Houston's decisive victory over Mexican president Santa Anna paved the way for Texas to become an independent republic.

A couple of years later, in 1838, the Army Corps of Topographical Engineers was founded to conduct the official reconnaissance, surveying and mapping duties for the US Army and government. This marked a major step forward in mapping the continental United States, and one of the Corps' key tasks was to explore potential routes for a transcontinental railroad line. Four courses were surveyed, and reports were filed with details of distances, grades, mountain passes, canyons, bridges and tunnels. Each survey also took into account the availability of timber, stone, coal and water. After the submission of a 13-volume report with numerous maps, charts and illustrations, the US Congress failed to choose one and the project lay fallow for years. However, when the Union Pacific–Central Pacific line was built after the Civil War, it followed one of the paths that had been proposed.

Perhaps the most accomplished of the topographical engineers was John C. Frémont, who led five exploring expeditions to the West between 1842 and 1854. In so doing, he earned such a reputation that in 1856 he was selected as the presidential candidate of the Republican Party.

Frémont's most significant expedition was his second, on which he blazed an overland route to Oregon that would replace the much longer course established by Lewis and Clark. Relying heavily on his wife's literary skills, Frémont produced a report about the journey which was so striking that the government printed thousands of copies for free distribution. Such was the popularity of the report and the accompanying map that they were released a further 25 times, and they became the essential guides for thousands of overland immigrants to Oregon and California.

The map was the first truly scientific one of the western United States. During the expedition, Frémont took thousands of longitude and latitude readings in order to obtain the most complete measurements possible of surface topography. The cartographer Charles Preuss then drew the map from Frémont's daily computations. Another landmark was that the map showed only country that had actually been traversed and measured by Frémont. As he wrote, "This map may have a meagre and skeleton appearance to the general eye, but is expected to be more valuable to science on that account, being wholly founded upon positive data and actual operations in the field ... almost every camping station being the scene of astronomical or barometrical observations."

The Frémont–Preuss map therefore featured an uncommon amount of "white space" and a relatively thin route plan, but it was the first to show an unbroken trail from the Great Plains to Oregon and California. Most importantly, it was entirely reliable – a hugely significant feature for a document that would be used by travellers to lead them through remote areas where the accuracy of reports relating to food or water could mean the difference between life and death.

Above: A photograph of John C. Frémont (1813–90), called "The Great Pathfinder". After his career as an explorer, he served as a United States Senator from California, and as Governor of the Arizona Territory.

Opposite: A map of Florida from the 1825 revised edition of Henry Tanner's classic *A New American Atlas*. The maps in Tanner's atlas were printed on high-quality paper and hand-coloured.

Overleaf: The 1845 Frémont–Preuss map of the territories explored on Frémont's expeditions of 1842 and 1843–44. Considered the finest map of the region that had ever been produced, it updated and expanded the one published two years earlier after the first expedition.

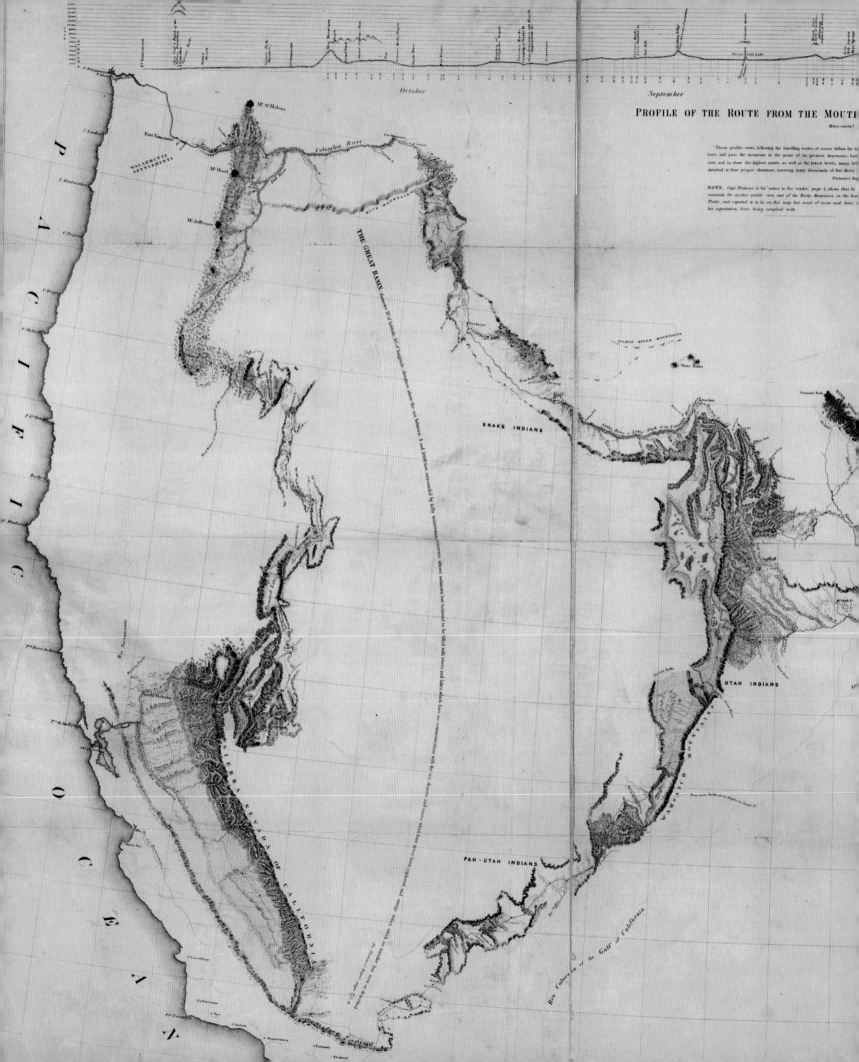

*These profile views, following the travelling routes of course follow the h...
lines, and pass the mountain at the point of its greatest depression; but...
view, and to show the highest points as well as the lowest levels, many lo...
sketched at their proper elevations, towering many thousands of feet above...

*Fremont's Re...*

NOTE. Capt Fremont in his notice to the reader, page I, shows that he h...
materials for another profile view, east of the Rocky Mountains on the ho...
Platte, and expected it to be on this map; but want of room and time...
his expectation from being complied with.

Great Salt Lake

P
A
C
I
F
I
C

O
C
E
A
N

Mt St Helens

Fort Vancouver

WALAHMUTTE
SETTLEMENTS

Mt Hood

Columbia River

BLUE MOUNTAINS

Mt Jefferson

THE GREAT BASIN. diameter II of latitude, 10 of longitude: elevation above the sea between 4 and 5000 feet: surrounded by lofty mountains: contents almost unknown: but believed to be filled with rivers and lakes which have no communication with the sea, deserts and oases which have never been explored: and only known through trappers...

SNAKE INDIANS

SALMON RIVER MOUNTAINS

Three Buttes

Fort Hall

Snake River or Lewis Fork of the Columbia

Fort Boisé

ROCKY

WIND

Fremont's Peak

SIERRA NEVADA OF CALIFORNIA

Pyramid Lake

Nueva Helvetia

Rio Sacramento

UTAH INDIANS

Sevier Lake

WAHSATCH MOUNTAIN

PAH-UTAH INDIANS

Rio Colorado or the Gulf of California

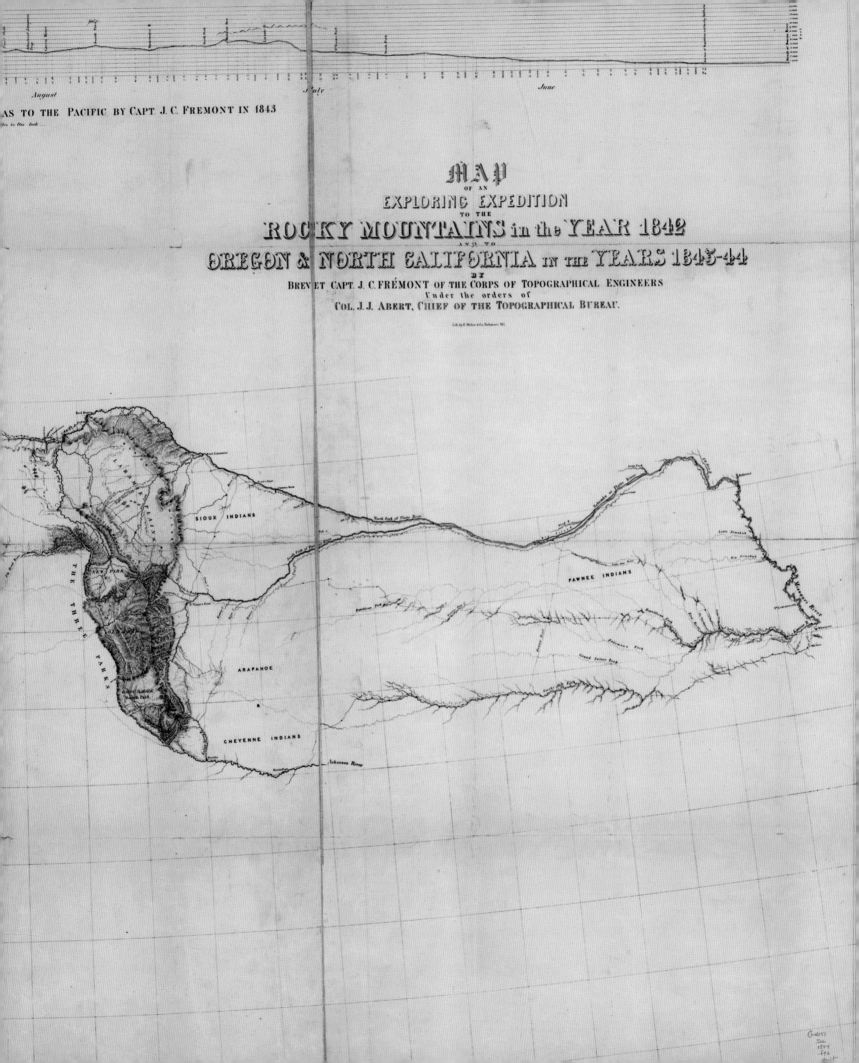

August  July  June

...AS TO THE PACIFIC BY CAPT. J.C. FREMONT IN 1843

...les to One Inch...

# MAP
## OF AN
### EXPLORING EXPEDITION
#### TO THE
## ROCKY MOUNTAINS in the YEAR 1842
#### AND TO
## OREGON & NORTH CALIFORNIA in the YEARS 1843-44
#### BY
#### BREVET CAPT. J. C. FRÉMONT OF THE CORPS OF TOPOGRAPHICAL ENGINEERS
#### Under the orders of
#### COL. J. J. ABERT, CHIEF OF THE TOPOGRAPHICAL BUREAU.

Lith. by E. Weber & Co. Baltimore Md.

SIOUX INDIANS

North Fork of Platte River

NEW PARK

THE THREE PARKS

PAWNEE INDIANS

ARAPAHOE

Grand Saline Fork

CHEYENNE INDIANS

Arkansas River

# Mapping the
# European Empires

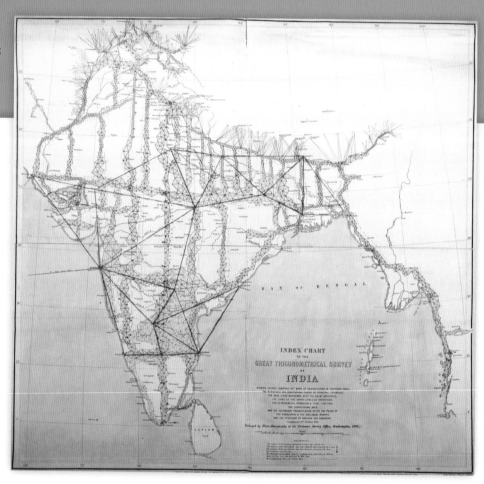

Left: The "Great Theodolite" used by William Lambton and then George Everest in the Great Trigonometrical Survey of India. It was so accurate that it helped produce the most precise maps of India that had yet been created.

Above: The Index Chart to the Great Trigonometrical Survey of India. Lambton's triangulation established the framework for an accurate map of the subcontinent and the measurements shown on the northern fringes allowed the first accurate height assessment of the Himalayas.

**W**HILE THE **U**NITED **S**TATES PUSHED INEXORABLY ACROSS THE CONTINENT of North America, European empires – now generally banished from that region – had to look in different directions to expand and consolidate. Accurate maps of the areas that their acquisitive eyes fell upon were increasingly important at a time when knowledge truly *could* mean power, control and wealth.

In 1798, five years before he sold Louisiana, Napoleon invaded Egypt. In a remarkably prescient act for later scholarship, he took with him 150 *savants* – scientists, engineers and historians – to investigate every conceivable aspect of Egypt's past. Their findings began to be issued in 1809, in the 23-volume *Description de l'Égypte*. Ironically, the part that was intended to be published first – the atlas of maps – appeared last, in 1828. This was because the French cartographic effort, with 37 engineers working under Pierre Jacotin, had produced a map of such an inordinately high detail and accuracy that Napoleon's commanders convinced him to keep it secret for military purposes. Not until Napoleon was exiled to Elba was the map finally sent for engraving – its 47 sheets covered every section of Egypt with a precision that made them appear more like satellite images than hand-drawn maps.

While the French were in Egypt, the British were planning the systematic triangulation and mapping of India. The work began in 1802, with Lieutenant William Lambton establishing an initial baseline measurement in Madras. Using the "Great Theodolite", which weighed more than half a ton and required a dozen men to carry, he completed a triangulation down the coast and across the southern part of the subcontinent before he began the measurement of a central arc, which in the next 21 years he extended more than 1,100 km (700 miles).

Lambton's operation was officially named the Great Trigonometrical Survey in 1818. He died in 1823 and was succeeded by George Everest, who continued the survey up the spine of India. Having completed the longest arc of the meridian determination ever made, 21°, Everest retired in 1843.

Under his successor, Andrew Waugh, the field triangulation was completed and the heights of 79 Himalayan peaks were determined – including Mount Everest (named for Waugh's predecessor) – although the data assessment continued for many years.

The trigonometrical data were not of sufficient interest for general publication, but throughout Lambton's directorship the British East India Company slowly shifted from restricting the flow of geographical information to allowing its distribution. The result was the production of the *Atlas of India*, published for the Society for the Diffusion of Useful Knowledge. From the recent definitive surveys, the regional maps were produced at four miles to the inch (approximately 2.5 km to the centimetre). In theory, the remarkable level of accuracy attained with these maps meant that each copper plate would never have to be corrected – something that

was the bane of map engravers – and would, instead, only need material to be added. The *Atlas* remained the primary cartographic venture of India throughout the nineteenth century.

Although they were not surveyed with the same scientific precision as India and the Himalayas, the regions farther north and west also became of increasing interest to the British government and the British East India Company. As

early as 1813 Russia had taken Azerbaijan and Daghestan from Persia. Thereafter, Russia's relentless encroachment on the strongholds of Kokand, Khiva and Bokhara led the British to fear that Afghanistan would be used as a staging point for an invasion of India. What became known as "The Great Game" – the strategic rivalry conducted in the hidden passes and mysterious valleys of Central Asia – prompted the publication of numerous maps throughout Europe,

Above: Adolf Stieler's map of Iran and Turan, from the *Stielers Handatlas* of 1875, published under the direction of August Petermann. The world's leading scientific atlas, it often used terms such as "Turan": the Middle Persian name for Central Asia.

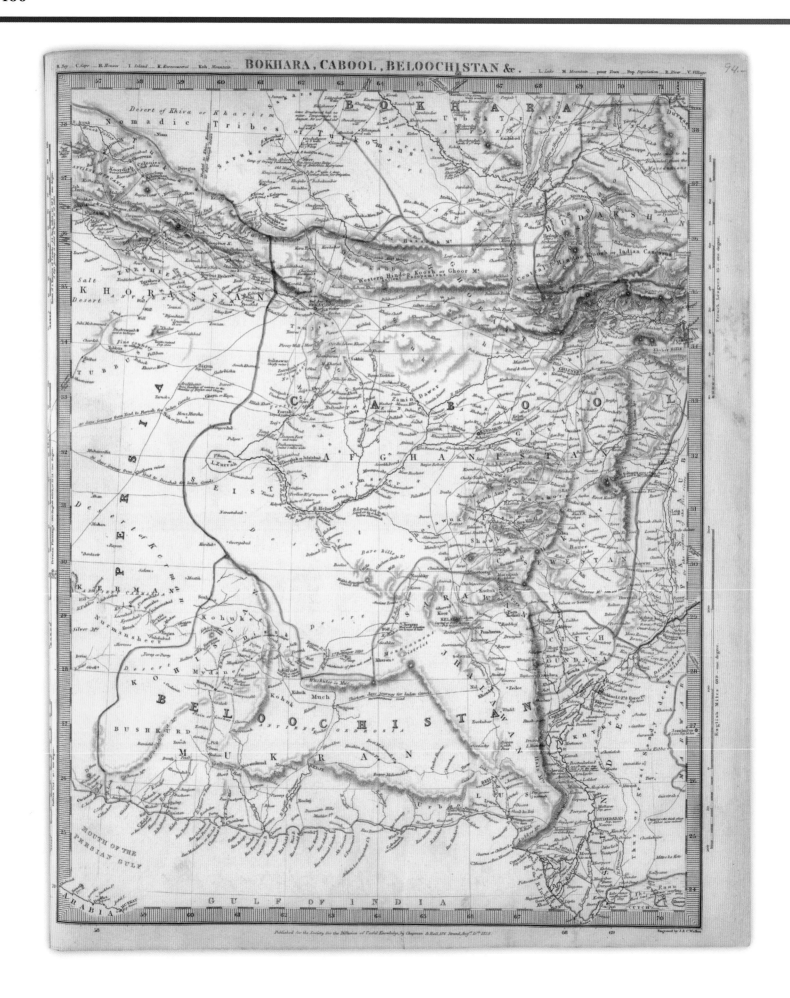

Published for the Society for the Diffusion of Useful Knowledge, by Chapman & Hall, 186 Strand, Aug.st 15th 1838.

which showed the implied threat to India and generated strong anti-Russian feelings among the British public.

As the Russian Empire expanded in one direction, it contracted elsewhere. Fearing its inability to defend its North American holdings, in 1867 Russia sold Alaska to the United States. "Seward's Folly", as the purchase was known by political opponents of US Secretary of State William Seward, created more than one and a half million sq km (nearly 600,000 sq miles) of new American territory at a cost of $7,200,000 – a rate of approximately five cents per hectare (two cents per acre). The US government immediately issued maps of the new territory, and within a few years there were many more.

Meanwhile, the successors to Napoleon – monarchist, imperial and republican – continued to oversee French colonial development. Beginning in 1830, the French conquered and then administered Algeria, which remained under French control until 1962. Victor Lavasseur's *Atlas National Illustre* – noted for the lavish illustrations that surrounded the maps – was typical in portraying Algeria not as a British-style colony, but as an integral French *département*.

The French also acquired colonies in Southeast Asia, gaining control in the 1860s of the ports of Annam and Tonkin and all of Cochinchina. Shortly thereafter, Cambodia became a French protectorate. Such developments were accompanied by updated maps, proudly showing these territories as French. One example was Adrien Brue's *Atlas Universel*, reprinted many times since it had first been published in 1822. The expansion of France, like that of Britain and Russia, would continue throughout the century.

Opposite: This map is taken from the *Atlas of India* and shows parts of what are now Uzbekistan, Afghanistan and Pakistan.

Below: Victor Lavasseur's 1856 map of Algérie Colonie Française from the *Atlas National Illustre*. This was one of the last atlases with such extensive and beautiful border decorations, which in this case tell much of the history of the region.

Overleaf: Based on the U.S. Coast Survey Office Survey produced by Charles Sumner, this map shows the territory (now known as Alaska) which was ceded by Russia to the United States. It was drawn by A. Lindenkohl and published in 1867.

SITKA
AND ITS APPROACHES
From Russian and British authorities.
Scale 200,000

Statute Miles
Nautical Miles

KRUZOV ISLAND
BARANOFF I?
SITKA SOUND

RUSSIAN AMERICA
A S S I A

SEA OF OKHOTSK
GHIJINSK BAY
Gulf of Penjinsk
Penshinsk
Ghijinsk

K A M T S C H A T K A

Cape Piughin
Cape Porotaith
Ukinska Bay
Karagnusky I.
New Kamtschatka
Cape Osernoy
Cape Kamtschatka
Gulf of Kamtschatka
Petropaulowski
Kronotzkoi Bay
Cape Kronozky
Cape Tschipunski

KOMANDORSKI I?S
Behring I.
Copper I.

C. Gowensky
Cape Olintorski
C. Omchinsky

C. St. Thaddeus
Bay of St. Gabriel

GULF OF ANADIR

Anadir Bay
Village
Holy Cross Bay

A R C T I C

Kolyma Bay
Ayan I.
Chaun Bay
Cape Shalagskii
Cape Yakan
Takokagui

T C H U K T C H E S   P E N A

Kolingchin Bay
East Cape
Diome

BEHRING STRAIT

Kelsegille
Kayne I.
C. Chapin
Cape Choukotski

St. Lawrence I.

Hall I.
St. Mathew
Pinnacle

Forty fathom curve

St. Paul   Russian Factory
PRIBYLOV ID.S
St. George   Factory

B E H R I N G   S E A

A L E U T I A N   I S L A N D S

Attou I.
Semichi I?s
Agattou I.
Tchitki I. or group
Bouldyr I.
Kyska
Tchegoula I?s
Semisopochnoi
Rat I.
Goreloi
Amtchitka I.
Kanaga I.
Tanaga I.
Youdak
Atkha
Sitchin
Amlia I.
Adakh I.

BLIJNIE or RAT ID?S

A N D R E A N O W S K Y   I D?S

Segouam Pass
Amoukhta Pass
Ids of the four Mountains
Younaska
Amoukhta
Segouam
Bogoslov Volcano
Oumnak Pass
Oumnak

N O R T H   P A C I F I C

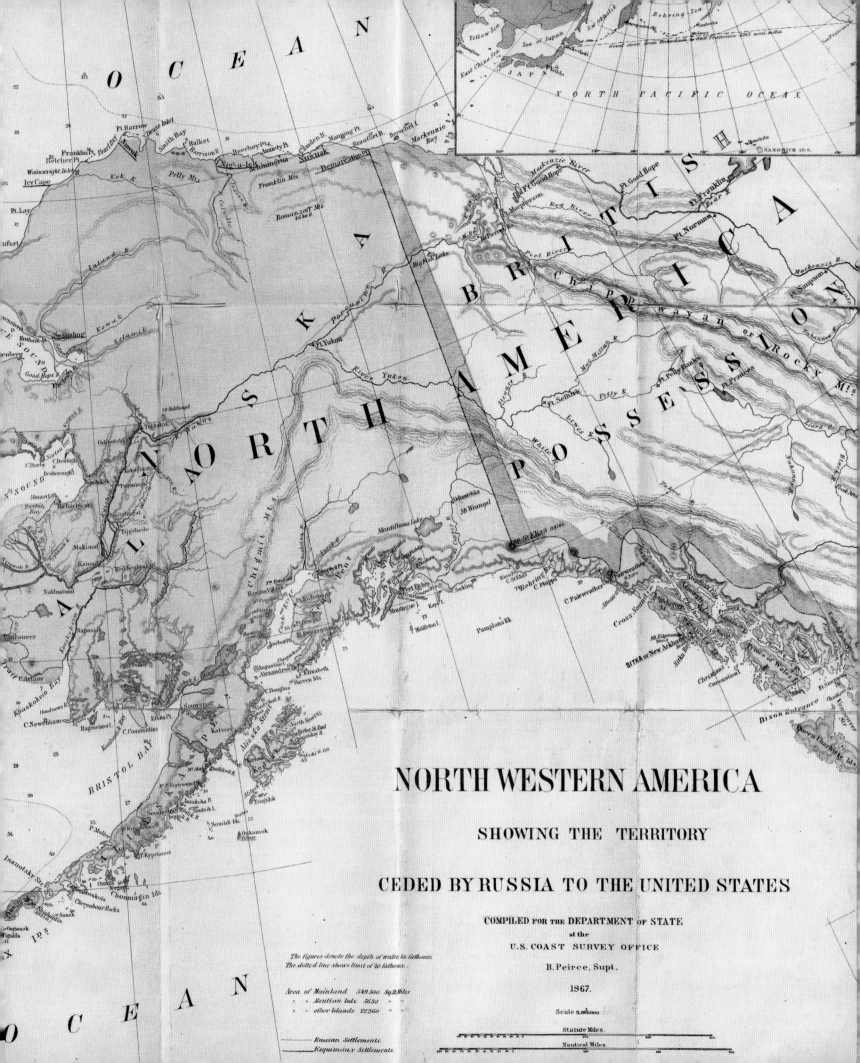

# NORTH WESTERN AMERICA

## SHOWING THE TERRITORY

## CEDED BY RUSSIA TO THE UNITED STATES

COMPILED FOR THE DEPARTMENT OF STATE

at the

U.S. COAST SURVEY OFFICE

B. Peirce, Supt.

1867.

Scale 5,000,000

Statute Miles.

Nautical Miles.

The figures denote the depth of water in fathoms.
The dotted line shows limit of 40 fathoms.

Area of Mainland   549,500 Sq.º Miles
  „   „ Aleutian Isds.   5630
  „   „ other Islands   22260

Russian Settlements.
Esquimaux Settlements.

# The Dark Continent

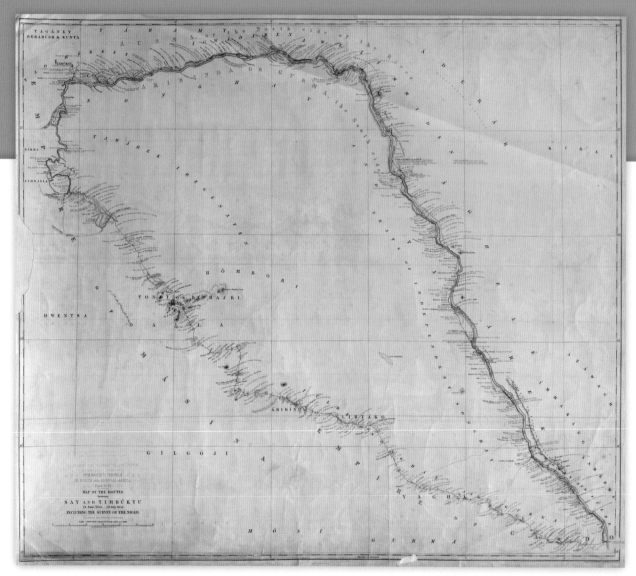

Left: Heinrich Barth's map of the Niger region, showing parts of his travels in 1850–55. Barth's five-volume *Travels and Discoveries in North and Central Africa* is a classic account of scientific exploration.

Below: David Livingstone, looking younger than he did in most portraits. Usually portrayed as craggy and worn-looking, he was actually only 52 years old when he disappeared for the final time into central Africa.

CONSIDERING THE MANY CENTURIES THAT EUROPEANS STRUGGLED to learn the location of the sources of the River Nile, the story of the accurate mapping of the interior of sub-Saharan Africa is surprisingly brief. In many senses the tale began in 1788 in London, with the formation of the Association for Promoting the Discovery of the Interior Parts of Africa. Although the outline of the continent had been delineated, the lack of knowledge about the geography of its interior was made clear by a map compiled for the association by James Rennell. Yet just a century later, Henry Morton Stanley had charted one of the continent's last-known major geographical features.

In its early years the association sponsored and encouraged many ventures to discover the source and course of the Niger river, or to reach the fabled city of Timbuktu near that river. Some of these expeditions began by crossing the Sahara Desert – such as the one mounted in the 1820s by a party that included Hugh Clapperton and Dixon Denham, which discovered and mapped Lake Chad and blazed a route to the city of Sokoto, near the Niger.

However, more explorers approached from the west, including Mungo Park, whose published account of following the Gambia river to the Niger was accompanied by a map from Rennell. Unfortunately, Park disappeared

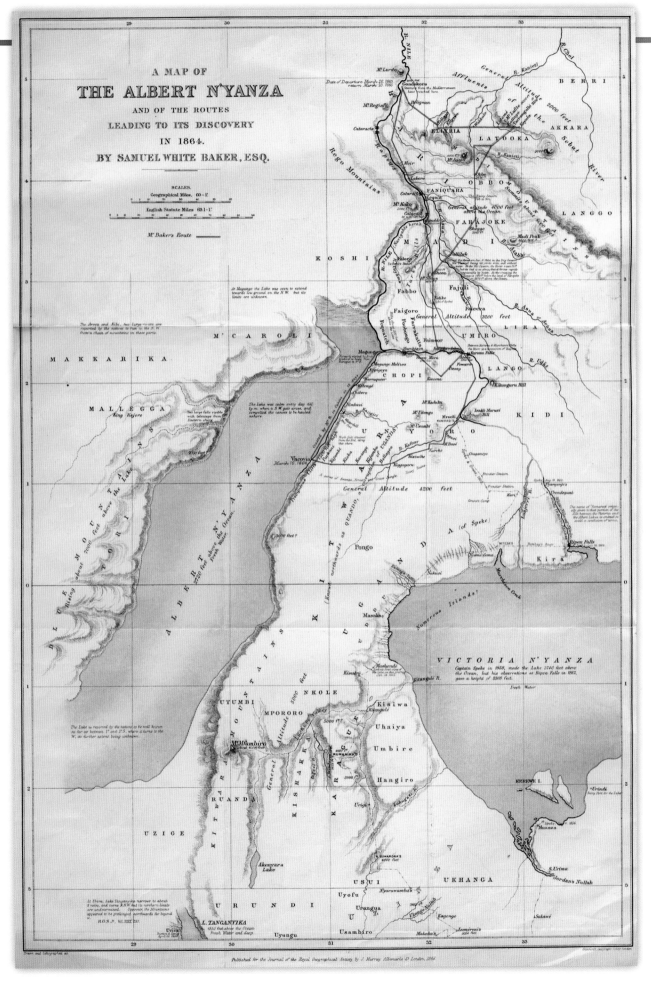

Left: Samuel Baker's map of the Albert N'yanza, today called Lake Albert. When Baker's party first reached the lake, he named it in honour of the late Prince Albert, consort of Queen Victoria.

Overleaf: A map of David Livingstone's route on his crossing of Africa during 1853–56. This appeared in his best-selling book *Missionary Travels and Researches in South Africa*.

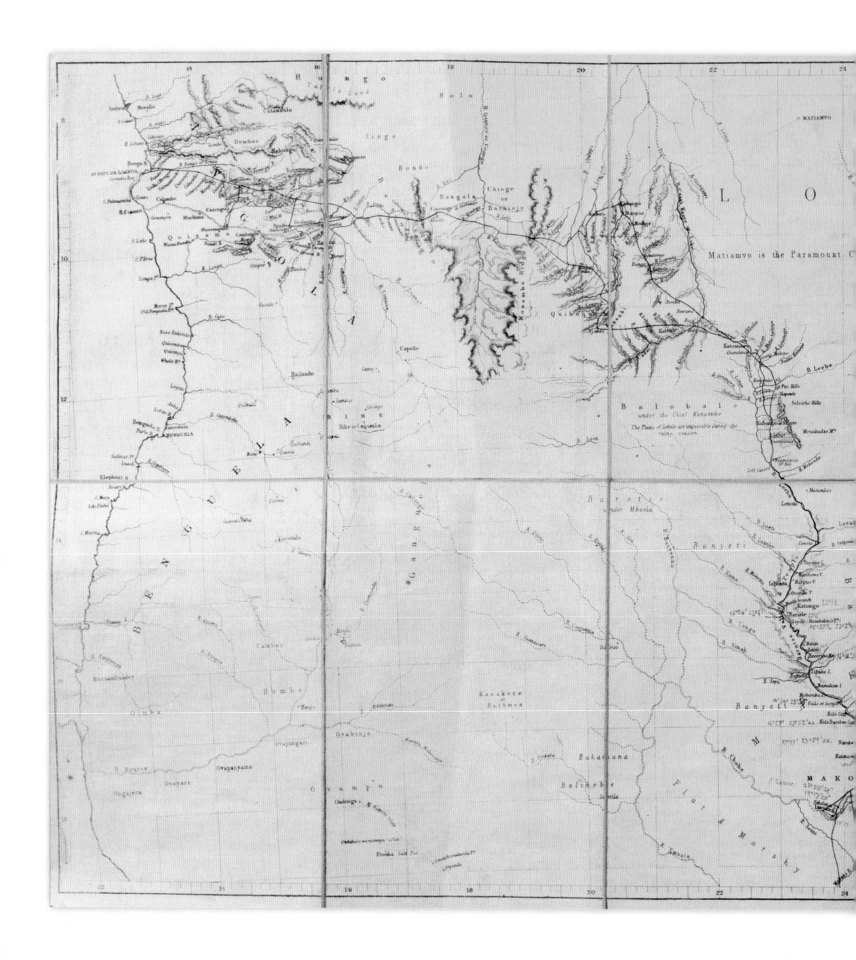

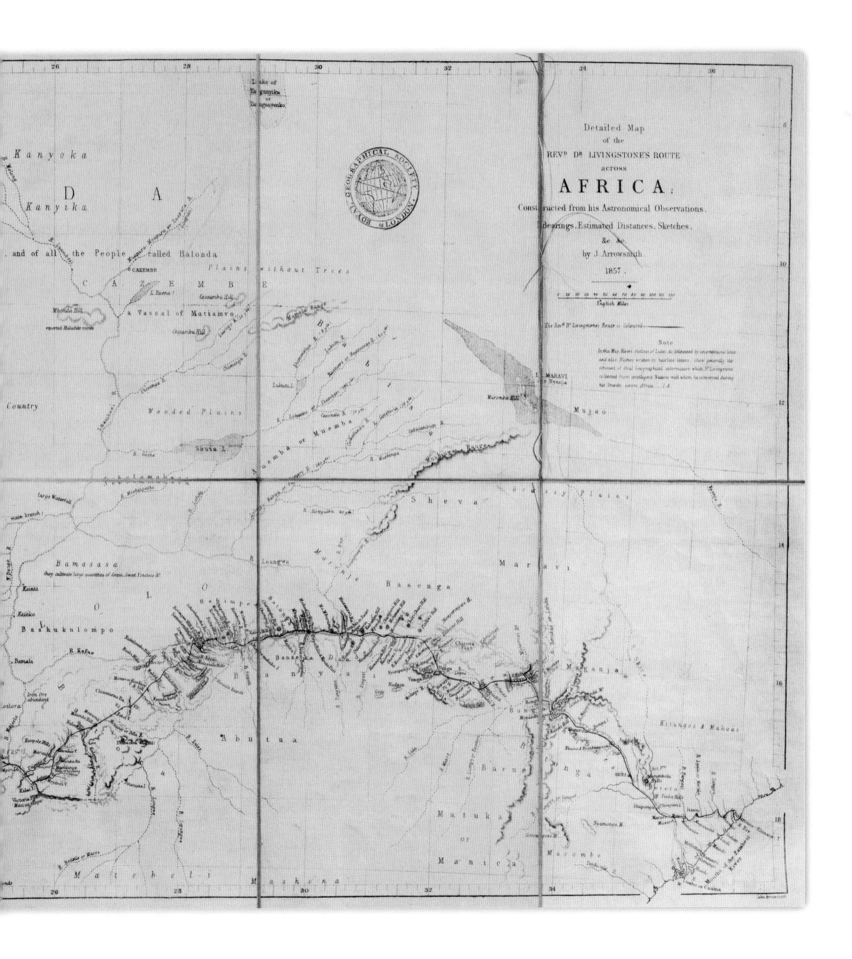

on his next expedition, and there was little successful exploration in the region until 1827–28, when a European finally reached Timbuktu and returned to tell the tale. René-August Caillié's travels – during which he posed as a displaced Egyptian – took him from the coast of Guinea to Djenné and thence downriver to Timbuktu. From there, he completed his 7,200-km (4,500-mile) journey by crossing the desert and passing through the Atlas Mountains en route to Tangier.

Meanwhile, in 1825 Clapperton made a return expedition, starting from the Bight of Benin with his manservant Richard Lander. They again reached Sokoto, but there Clapperton died. Lander published an account of their travels, and then in 1830 he led his own expedition, accompanied by his brother John. The pair set off from Benin, reached the Niger river at Bussa, where Park had disappeared, and then sailed south, becoming the first Europeans to follow the route of the great watercourse to its delta.

Two decades later, Heinrich Barth re-explored much of the area covered by Clapperton, Caillié and the Landers. To their maps, he added his own in a five-volume tome. His account was the most scientific description yet of the Sahara and the Niger, enlarging upon many of the major geographical details of those regions.

At the far end of the continent, the Dutch and then the British had slowly moved inland from Cape Town. However, the greatest expansion of geographical knowledge about southern Africa came from the journeys of the medical missionary David Livingstone. In 1849 he crossed the Kalahari Desert and discovered Lake Ngami;

later, he went north and reached the Zambezi river. In 1853–56 Livingstone made the first crossing of Africa – negotiating the mountains of Angola, tracing most of the course of the Zambezi and discovering the falls he named in honour of Queen Victoria. Those travels made him a national hero in Britain, and with the sponsorship of the Royal Geographical Society (RGS) Livingstone returned to Africa to navigate the Zambezi in full. When this proved impossible, he explored the Shire river in an attempt to reach Lake Nyasa. Although this effort was generally considered a failure, Livingstone's maps of the two expeditions greatly clarified the river systems and general geography of southern Africa.

In the period overlapping Livingstone's two expeditions, the search for the sources of the Nile took on a new immediacy. Ptolemy had long before produced maps that showed a range of mountains draining into a pair of lakes, out of which flowed the Nile. When Ptolemy's *Geographia* was translated into Latin, the lakes and the "Mountains of the Moon", as they were called, were introduced to European cartography and they became an enduring feature of maps of Africa for hundreds of years. Even John Tallis's *Illustrated Atlas and Modern History of the World* – released to coincide with the Great Exhibition of 1851 and considered a cartographic masterpiece – recorded these as-yet-undiscovered mountains.

In June 1857, the British explorer Richard Francis

Above: A portrait of John Hanning Speke, who died in a hunting accident at the age of 37, the day he was scheduled to hold a public debate with Richard Burton.

Below: Speke's 1858 painting of the settlement of Mgongo Thembo. According to Henry Morton Stanley, the name means "elephant's back", referring to the rocks resembling that vast creature.

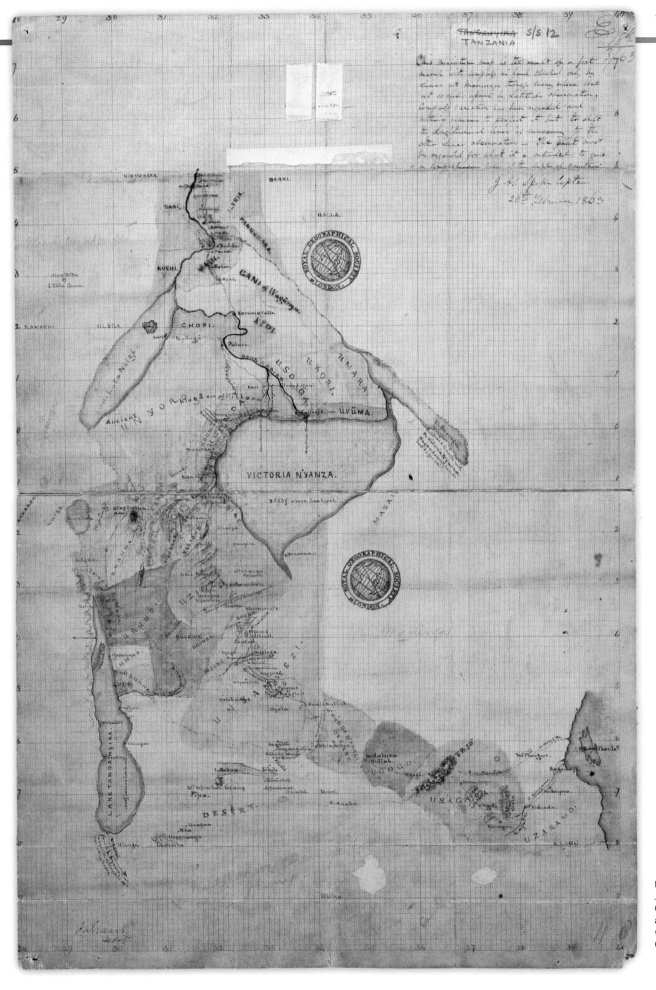

Left: A map showing the route of John Hanning Speke and James Grant from Zanzibar to Lake Victoria and north to Gondokoro. The map was drawn and hand-coloured by Grant in 1863.

Burton – already famous for remarkable journeys to Mecca, Medina and Harar – left Zanzibar in search of a vast lake rumoured to be far inland. He was joined by John Hanning Speke, an officer in the Indian Army. In February 1858, the pair discovered Lake Tanganyika, but due to ill health they were unable to study it extensively.

On their return to the coast, Speke made an excursion north and discovered another great lake that he claimed – without any evidence – was the source of the Nile. He and Burton disagreed vehemently about whether Lake Tanganyika or Lake Victoria was the Nile's source, but in 1860 the RGS sided with Speke, sending him and James Grant to continue the investigation. Although impressive, their journey was not entirely successful because Speke's map clearly showed that at points they left the course of the Nile and therefore could not prove with certainty that they had followed the same river.

As Speke and Grant travelled down the Nile, they met a party led by Samuel Baker, which included Baker's wife Florence. During the next two years, the Bakers found another great lake to the west of Speke's route, which they named Lake Albert. Although the Bakers showed that the waters of the Nile flowed into and back out of Lake Albert, they, too, left sections of the river unexplored. Their map proved that the region still needed a full cartographic survey.

Hoping to resolve the question of the sources of the Nile once and for all, in 1866 the RGS sent David Livingstone back to Africa. However, although the travel journals he compiled during the next seven years helped to make him one of the great Victorian icons, Livingstone did not hugely further geographic understanding of the region. That was left to two men who drew their inspiration from him.

## The Other Source of the Nile

**Before the mid-nineteenth century race to discover the sources of the Nile, the search had actually, in part, already been successful. Far south of where the Nile provides the lifeline for Egypt, beyond the six great cataracts, Khartoum sprawls at the confluence of two rivers: the White Nile, flowing north from Lake Victoria, and the Blue Nile, dropping from the highlands of Ethiopia. The latter, which contributes about 80 per cent of the water volume, flows approximately 1,450 km (900 miles) from Lake Tana, itself fed by several streams. The first European to follow the primary stream to the springs at Gish Abay – reaching the ultimate source of the Blue Nile – was Pedro Paez, a Spanish Jesuit missionary, who did so in 1618. A decade later, a Portuguese Jesuit, Jerónimo Lobo, repeated the feat, and then in 1770 the Scottish explorer James "Abyssinian" Bruce, whose accurate tales were ridiculed as being too extraordinary to be true, did the same.**

In 1871 Henry Morton Stanley – a reporter for the *New York Herald*, one of the world's most prominent and enterprising newspapers – located the "missing" Livingstone on the shores of Lake Tanganyika. The reports of their meeting made Stanley an international figure and put Livingstone back in the limelight. In the expedition's aftermath, in 1873 the British traveller William Winwood Reade produced a "Map of African Literature", which featured the names of different explorers in the regions they had investigated.

After Stanley's return, the RGS sent an expedition to assist Livingstone, selecting as its leader a Royal Navy officer, Verney Lovett Cameron. While he was recovering from fever in the interior, Cameron learned that Livingstone had died. Deciding to continue the explorer's work, he made a survey of Lake Tanganyika and then followed an outlet west to the mysterious Lualaba river. Cameron could not obtain boats to travel north, so he headed southwest into Katanga. In November 1875 he arrived at the Atlantic Ocean in Angola, having made the first crossing of Central Africa. His subsequent book and maps earned him the Founder's Medal, awarded by the RGS for "the promotion of geographical science and discovery".

While Cameron was in the field, Stanley also decided to carry on Livingstone's work. In 1874 he sailed around Lake Victoria in a prefabricated boat, confirming it to be a single lake that discharged solely at Ripon Falls – the beginning of the Nile. He then conducted measurements that showed Lake Tanganyika was not related to the Nile. Finally, he followed the Lualaba north into the Congo river, which he navigated to the coast, reaching the Atlantic 999 days after his departure. Stanley had resolved all the last great questions about the watersheds of Central Africa, and his two-part map of the journey in *Through the Dark Continent* (1878) remains a classic for its accuracy, having been based on

Left: A photograph of Henry Morton Stanley. Stanley's crossing of Africa (1874–77) – sponsored by the *New York Herald* and the *Daily Telegraph* – was perhaps the most successful of all expeditions to the "Dark Continent".

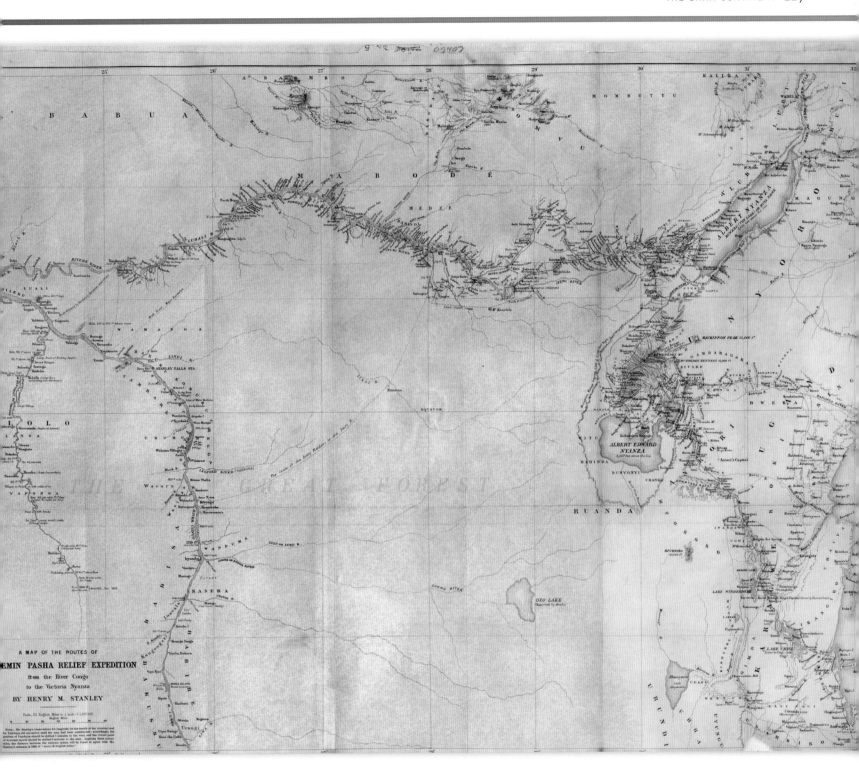

A MAP OF THE ROUTES OF
EMIN PASHA RELIEF EXPEDITION
from the River Congo
to the Victoria Nyanza
BY HENRY M. STANLEY

scientific observations taken under the most difficult and dangerous of conditions.

Stanley's last great cartographic success came on the Emin Pasha Relief Expedition (1887–89), an effort to rescue members of a military outpost at Lake Albert. Stanley crossed Africa via the Congo rather than from the coast of East Africa, and the passage through the unknown Ituri rainforest was perhaps the most dreadful ordeal any explorer of Africa ever faced. After six months, during which more than half of the party died, they reached Lake Albert. On his subsequent trek eastward, Stanley made his last discovery: the Ruwenzori Range (now known as the Rwenzori Mountains). When these mountains were placed on the two major route maps in Stanley's expedition account, this completed the link between the heart of Africa and the Nile, for it is these peaks that ever since have been thought of as Ptolemy's "Mountains of the Moon".

Above: A map showing the routes of Henry Morton Stanley and the Emin Pasha Relief Expedition on the horrific treks through the Ituri rain forest to Lake Albert. Stanley then led Emin Pasha's party to the east coast, sighting the Ruwenzori Range *en route*.

# Mapping a New Continent

Left: Matthew Flinders's General Chart of Terra Australis or Australia, based to a great extent on his own masterful voyages and charting of the continent's coastline.

Below left: A drafting pen, compass dividers and pocket telescope belonging to Matthew Flinders. He produced such fine maps that his work has been compared to that of James Cook and George Vancouver.

Opposite: An early chart of the interior of New South Wales, drawn by John Oxley. This version was released in 1825 by the sons of the famed London cartographer Aaron Arrowsmith.

THE DUTCH EAST INDIA COMPANY HAD ALREADY BEEN PRODUCING MAPS of New Holland for about 150 years when James Cook landed at Botany Bay in 1770. But within two decades, it was the British, not the Dutch, who had established a colony. Almost immediately after the arrival of the "First Fleet" in 1788, a survey was made of the fine natural harbour called Port Jackson and a map of it produced by 19-year-old midshipman George Raper.

Settling in New South Wales proved harder than foreseen, so it was several decades before the interior was investigated. Meanwhile, the continent's outline was clearly delineated. In 1798–99, Matthew Flinders and George Bass, having already explored the coast of New South Wales, circumnavigated Van Diemen's Land, thus proving it to be an island rather than a peninsula.

Then, in 1801, Flinders began a thorough charting of the coastline, beginning at Cape Leeuwin in southwest Australia. The next year he met and shared his findings with Nicolas Baudin, who was conducting a competing French survey. Flinders mapped most of the coast of Australia before he had to abandon the task after his ship became unseaworthy. Making for England, he put in at Mauritius for repairs but he was accused of spying and detained by the French authorities for nearly seven years. Flinders died in 1814, the day after the account of his expedition, *Voyage to Terra Australis*, and its accompanying atlas were published.

Baudin had also spent two years on his survey. He died in Mauritius, three months before Flinders arrived there, during a return journey to France. The atlas for Baudin's expedition was prepared by Louis de Freycinet, who ignored the contributions of both Baudin and Flinders, claiming most of the credit for himself. The map of Australia appeared in 1811 and was the first to show the coast virtually complete.

The earliest successful exploration of the interior of Australia was a crossing of the Blue Mountains by Gregory Blaxland in 1813, but the map he made

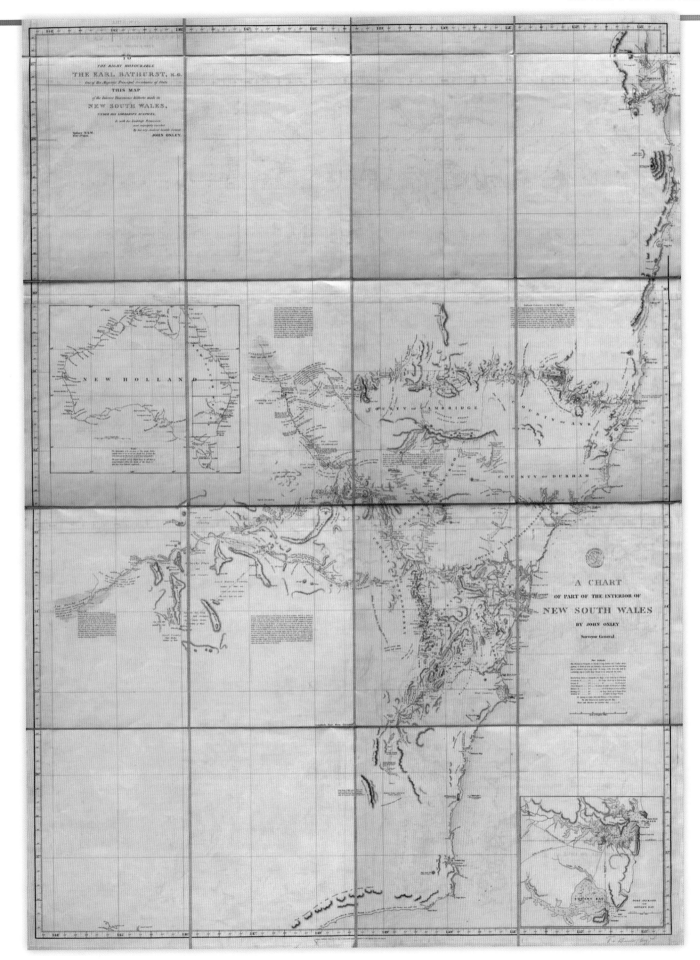

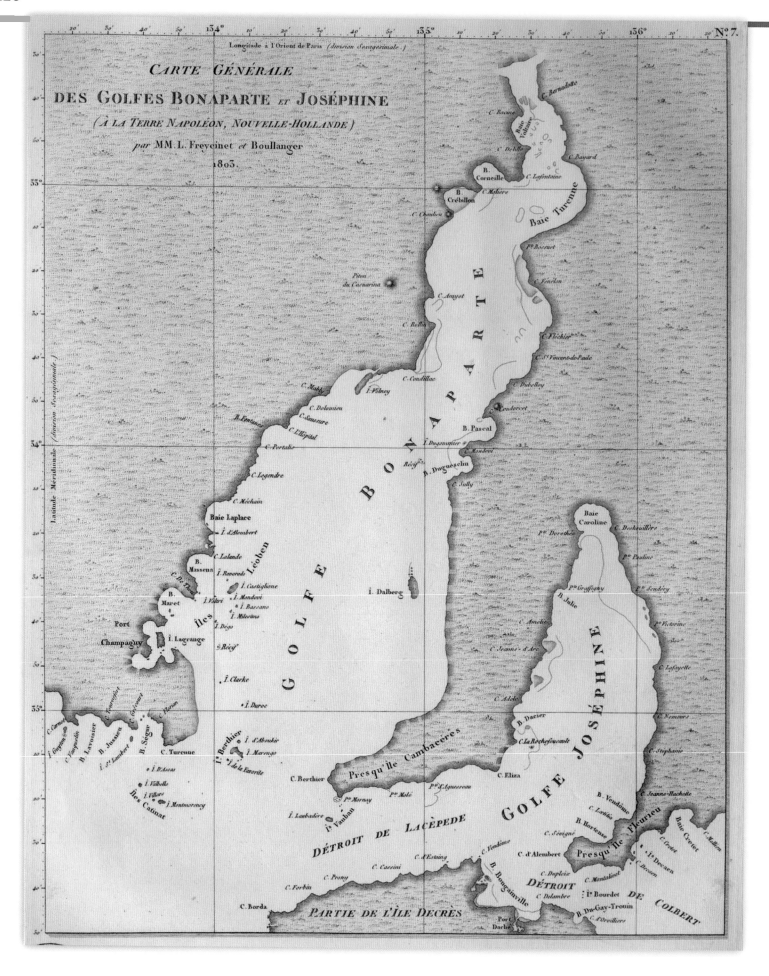

was strictly a route-finder. The first significant cartographic representation of the inland of New South Wales was produced by John Oxley, the colony's surveyor-general. Due to it being published by Aaron Arrowsmith, Oxley's map was guaranteed to receive international circulation when released in 1822.

During the decades that followed, maps of New South Wales, its individual counties and the whole of southeast Australia became more and more complete as Thomas Mitchell, Oxley's successor, gained details from his own expeditions and from those of Allan Cunningham, Charles Sturt and Edward Eyre.

Sturt left Adelaide on his fourth major expedition in 1844, hoping to find a great inland sea and a mountain range at the heart of the continent. He failed to do so, but travelled closer to the centre of Australia than had been done before. His map of the journeys was produced by John Arrowsmith and included descriptions of previously unknown topography and vegetation.

While Sturt was in central Australia, Ludwig Leichhardt, a German immigrant, headed northwest from Brisbane

## The Great Inland Sea

**One belief held by many explorers of Australia was that hidden somewhere was a large river system, as was the case in other continents. When this was not discovered, some replaced it with the concept of a great inland sea. Thomas Maslen, a former British East India Company officer, had one of the clearest – if most outlandish – images of the interior. His book _The Friend of Australia_ was published in London in 1830, and it included a map that was borne out of little other than his imagination, with Australia almost divided in two by a large inland sea connected to a river system that flowed into the Indian Ocean.**

toward Port Essington in the extreme north, reaching it in December 1845, despite being a poor navigator. Leichhardt's map was drawn by Samuel Perry, Mitchell's deputy, with two interesting additions: an illustration of Leichhardt and the indication of north shown by a bow and arrow.

Two expeditions attempted to cross the continent from south to north in 1860. John McDouall Stuart, already known for opening the region around Lake Eyre, pushed about 1,300 km (800 miles) through unexplored territory before he was forced back by a lack of supplies and a confrontation with Aborigines. Stuart's map was printed in Adelaide and in Germany by the geographer August Petermann, but it did not receive much other interest.

This lack of attention was due to the fanfare that accompanied another party, which went north from Melbourne. Unfortunately, the expedition leader, a policeman named Robert O'Hara Burke, failed to tell the main party of his definitive plans before he dashed ahead with three others, who included his deputy, William Wills. Thus, on their return (having retreated before reaching the Gulf of Carpentaria), no supplies were to be found. Three of the men died and the fourth was only saved after being cared for by Aborigines.

This tragedy received extensive attention worldwide, virtually to the exclusion of Stuart, who successfully completed his continental crossing in 1862. Numerous maps showed the route of Burke and Wills, whereas Stuart's map – of a successful expedition – was never widely reproduced.

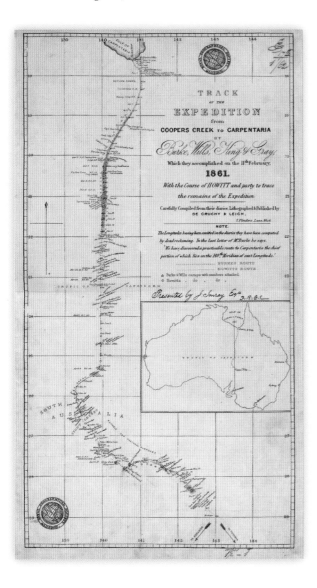

Opposite: A map showing the French names given to parts of the Australian coast on Nicolas Baudin's expedition of 1800–03. The maps were produced by Louis de Freycinet.

Left: A map of the route of the trans-Australia expedition led by Robert O'Hara Burke and William Wills. The final party of four was halted by mangrove swamps only a few miles from the Gulf of Carpentaria, and three of them tragically perished on the return journey.

Below: Drawings of wombats from _Voyage de découvertes aux Terres Australes_, the expedition account by François Péron and Louis de Freycinet. The Australian marsupials were highly unusual to Europeans at the time.

# Sven Hedin
## and Central Asia

Left: The most influential modern explorer of the remote and previously little-known region in the heart of the Eurasian landmass, Hedin was able to make use of his linguistic skills and ability to interact with the local people as well as his extensive scientific knowledge and cartographic ability.

I N THE HISTORY OF EXPLORATION perhaps no other man has reported on a region of the world in such depth and breadth as the Swedish scientist, cartographer and artist Sven Hedin. Between 1893 and 1935, he explored, investigated and charted lands never before seen by Europeans, while conducting four major expeditions that criss-crossed Tibet, the Himalayas and what was then known as Chinese Turkestan (modern-day Xinjiang).

Hedin surveyed an entire range of previously unrecorded mountains, measured the flow and charted the routes of rivers, and correctly predicted hydrological changes that affected the position of "wandering lakes". He discovered, and then conducted excavations at, lost archaeological sites. He recorded pioneering ethnographic information, discovered valuable mineral reserves and demonstrated the geological diversity of unexplored areas. Throughout, he mapped locations and routes with a precision rarely seen.

Born in Stockholm in 1865, Hedin wanted to become an explorer from a young age, and his linguistic, artistic and cartographic talents placed him in good stead for this, as did his intellect, determination and ambition. After having met Hedin in Kashgar in Chinese Turkestan in 1890, the British explorer Francis Younghusband wrote: "He impressed me as being of the true stamp for exploration – physically robust, genial, even-tempered, cool and persevering ... I envied him his linguistic ability, his knowledge of scientific subjects, and his artistic accomplishments. He seemed to possess every qualification of a scientific traveller."

The young Swede's first major cartographic effort was a six-volume world atlas that he completed at the age of 18. Three years later he made the first of two journeys to Persia, Mesopotamia and the old Central Asian khanates of Bokhara, Samarkand and Tashkent. Fascinated, he prepared for future scholarly investigations by earning a doctorate under the renowned German geographer Ferdinand von Richtofen.

Hedin began his first major expedition in 1893, when he crossed the Pamirs – called "the Roof of the World" – east to Kashgar. Two years later, having barely survived the Taklamakan Desert, he entered it again and found, buried in the sand, ruins of ancient Buddhist settlements that had been outposts along the Silk Road. He continued to explore the Tarim Basin and then investigated the northern reaches of Tibet.

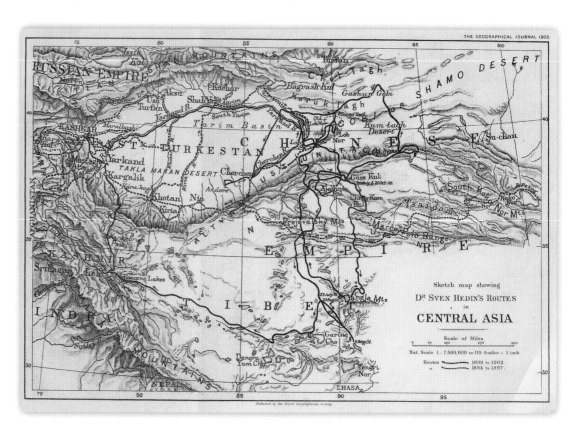

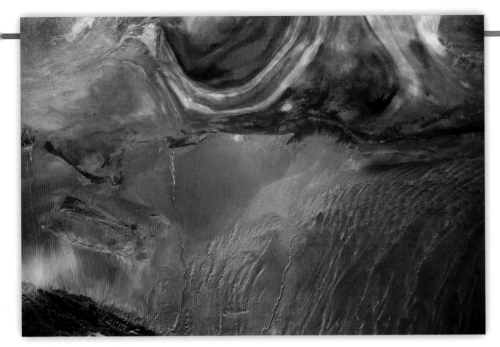

Opposite below: A map of Hedin's journeys during his first two major expeditions to Central Asia. Hedin produced his own maps from his detailed scientific observations and added to them with drawings of individuals and places.

Left: A view of the Lop Nor basin from the space shuttle *Endeavour*. The concentric rings were formed as the water in the now-dry lake disappeared in stages. The region has since been used by the Chinese government for tests of nuclear weapons.

Below: A map produced by Hedin on his second expedition into Chinese Turkestan, on which he mapped the course and surrounding lands of the Yarkand-daria, the river running to the north of the Taklamakan Desert.

Throughout his career, Hedin had travelled only with native assistants, but this changed on his last undertaking – the Sino-Swedish Scientific Expedition (1926–35). Started as a step in helping to establish a Germany–China air route, the expedition expanded into a broad scientific programme with Swedish, German and Chinese scholars. Hedin served as field overseer, diplomat and fund-raiser. The expedition into east Turkestan and Inner Mongolia produced 54 volumes of scientific reports in nine categories: geography, geology, paleobotany, invertebrate paleontology, archaeology, ethnography, meteorology, zoology and botany.

Hedin, who died in 1952, was one of the world's most honoured explorers, including being the last Swede to be elevated to the nobility. The five volumes of the *Central Asia Atlas*, produced to accompany his final scientific reports, were published posthumously and constitute a fitting tribute to the man who produced the first precise maps of the Pamirs, the Taklamakan, the Silk Road, Tibet and the Himalayas.

In 1898 Hedin published *Through Asia*, a two-volume account that made him a household name throughout Europe. He also released a scientific report, illustrated by detailed maps produced from 552 sketch maps he had made during his journey, which traced his 10,500-km (6,525-mile) route. These were the first scientific maps of Central Asia and had been made with the help of hundreds of measurements of latitude, longitude and altitude.

Hedin returned to Chinese Turkestan in 1899 and mapped the Yarkand-daria, the main river north of the Taklamakan. He investigated Lop Nor – a lake that he eventually proved had changed position through the centuries – and then he made his greatest archaeological find: the fabled city of Lou-lan. Hoping to reach Lhasa, he ascended to the high Tibetan plateau, but five days from his destination he was ordered to turn back and leave the Hidden Kingdom. Instead, he continued west through Tibet and eventually crossed into India. Hedin again published both a popular account and scientific results. His four-volume scientific report included 217 full-page photographs, 1,588 smaller photos or drawings – both pencil and watercolour – and 75 maps, which were made from more than 1,100 sketch maps produced in the field.

Hedin's third expedition (1906–08) again took him to Tibet, as well as to the mountain chain north of the Himalayas, which he called the Trans-Himalaya and was the first person to map. He also charted the sources of the Brahmaputra and Indus rivers. His subsequent nine-volume report included what is still the most complete survey of the region's physical geography, including 234 maps. It also contained geological, meteorological, astronomical and botanical analyses by other experts on Hedin's collections and data. In addition, he produced a three-volume atlas that included maps sheets of Central Asia, Tibet, Chinese Turkestan and the Pamirs; maps of his routes; 52 hypsometric maps (giving elevations by contours); and 552 hand-drawn panoramas.

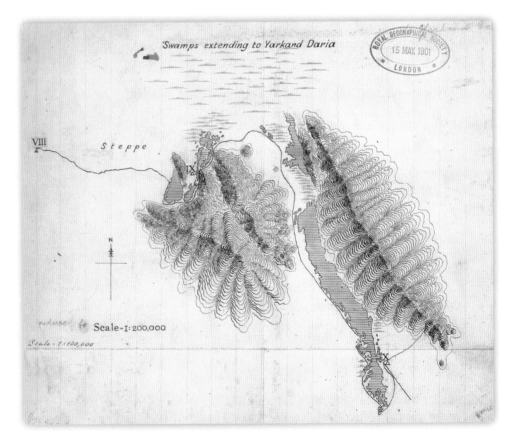

# Mapping the Oceans

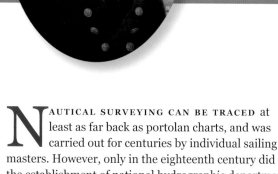

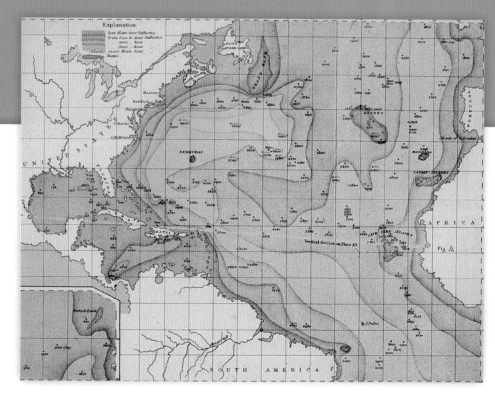

**N**AUTICAL SURVEYING CAN BE TRACED at least as far back as portolan charts, and was carried out for centuries by individual sailing masters. However, only in the eighteenth century did the establishment of national hydrographic departments – first by France (1720) and then Denmark and Britain – allow large coordinated programmes to survey, chart and describe the physical features and other factors that affected navigational safety.

Soon after the appointment of Britain's first Hydrographer of the Admiralty (see page 55) in 1795, the initial Admiralty charts were published. This series delineated coastlines with high and low water marks; recorded the depth of water as determined by soundings; noted navigational aids such as lights, buoys and beacons; and recorded geographical features that could pose hazards to shipping, such as reefs, shoals, hidden rocks and wrecks.

Such material proved so significant that in 1807 President Thomas Jefferson established the Coast Survey as the United States's first scientific agency. Run entirely as a civilian enterprise, its field programme began slowly, with the first survey only being made in 1834 along the south shore of Long Island. The first nautical chart did not appear until 1839. But when Alexander Bache, the great-grandson of Benjamin Franklin, was appointed director, operations sped up, and within a decade the Atlantic, Pacific and Gulf of Mexico coasts were being surveyed.

Meanwhile, Thomas Hurd, the second Hydrographer of the Admiralty, initiated what would become an international programme of maritime surveying by establishing a Corps of Surveying Officers in 1817 and ordering new vessels to be built. Under his successors – particularly Francis Beaufort – hydrographic surveys were carried out worldwide. The most famous of these was the second voyage of HMS *Beagle*, under Robert Fitzroy. Not only were standard coastal surveys made, but also observations of tides, currents, depths and the relationship of temperatures throughout the water column to weather and local fauna. Ultimately most significant, however, was the participation of Fitzroy's naturalist and geologist, Charles Darwin, who collected much of the data and made many of the observations that eventually led to his theory of evolution by natural selection.

Concurrently, there was a growth of interest in determining the ocean depths. Early bathymetric soundings, to calculate the depth of the water and thereby determine the topography of the ocean floor, were made by lowering a weighted line overboard until it reached the bottom and then measuring it. However, it could be difficult to tell when the line actually touched the seabed, and the ship could drift, thereby exaggerating the reading. Nevertheless, such techniques were used by both Charles Wilkes's US Exploring Expedition and Jules-Sébastien-César Dumont d'Urville's French expedition, with sounding locations determined by the use of a sextant and a chronometer.

Above left: A portrait of Matthew Fontaine Maury, the "Pathfinder of the Seas". A Virginian, Maury resigned his naval commission at the start of the American Civil War to serve the Confederacy.

Above: Matthew Fontaine Maury's map Basin of the North Atlantic Ocean, showing the bathymetry of that ocean, from his *Physical Geography of the Sea*.

Opposite: A map commemorating the completion of the trans-Atlantic cable in 1858. The first message was sent on 16 August, but the next month the cable failed.

The first serious effort to chart the bathymetry of the oceans was made by Matthew Fontaine Maury, superintendent of the US Depot of Charts and Instruments (which later became the US Naval Observatory) from 1842 to 1861. As had James Rennell before him, Maury studied old logs and charts and enlisted the help of sea captains to compile data he could use as the basis for a broad theoretical understanding of the oceans. This included information on winds, storms, surface currents, whale migration, chemical composition and the sea floor, all of which were discussed – and many mapped – in his classic book *The Physical Geography of the Sea*.

Maury also advocated bathymetric surveys using a deep-sea sounding device that had been created in 1852 by his colleague John Brooke. The next year, this was used aboard USS *Dolphin* when the first soundings were made on the Mid-Atlantic Ridge. This new finding gave birth to

Maury's notion of a "Telegraphic Plateau", along which could be placed a submarine trans-Atlantic telegraph cable. Although Maury did not actually have enough sounding data to qualify this thesis or construct valid bathymetric maps, his conclusions were of great interest to entrepreneurs keen to extend the range of the telegraph.

After several false starts in the years that followed (the first cable broke in the depths of the Atlantic), cables were successfully laid by 1866 and began carrying transmissions between continents. By 1928 there were 21 cables crossing the Atlantic and following a number of routes, each of which had required a series of soundings. Many soundings had also been taken on cable routes not ultimately used, such as that overseen by the Arctic explorer Francis Leopold McClintock, from Scotland to the Faroe Islands, Iceland, Greenland and Labrador. The depths of large areas of the Atlantic seabed had thus been investigated for the first time.

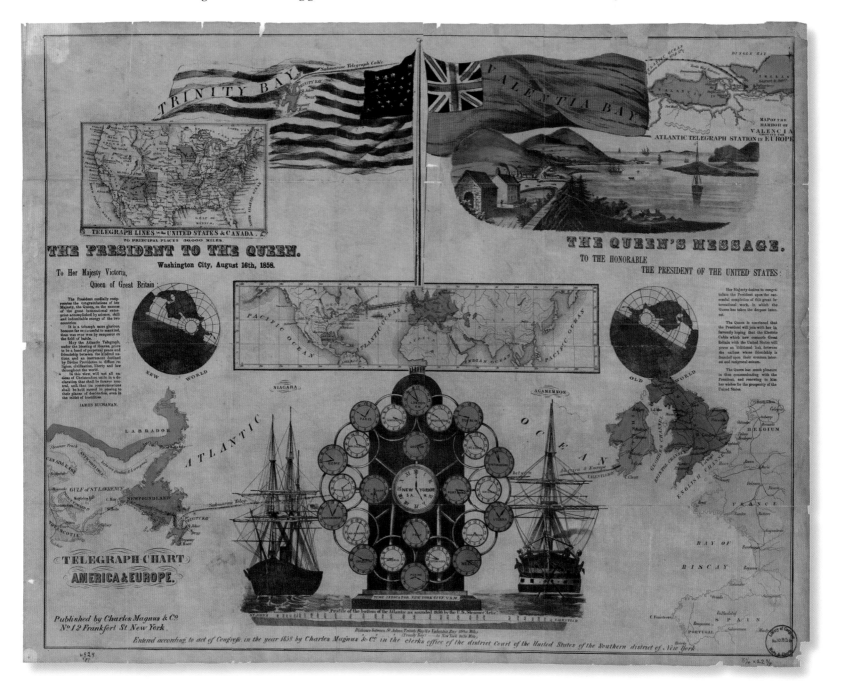

Petermann's Geographische Mittheilungen.

Warme Meeres-Strömungen.
Kalte Meeres-Strömungen.

Die Abstufung in den Farbentönen drückt die Mächtigkeit,
Schnelligkeit und Beständigkeit der Strömungen, die
Pfeile ihre durchschnittliche Richtung aus.

Die sich kreuzenden Strömungen sind mehr periodi-
scher Art, und die eine oder andere in gewissen
Jahreszeiten die vorherrschende.

Die zu den arktischen Strömungen gehöri-
gen Flussgebiete des Mackenzie und der
Sibirischen Flüsse, ferner das Becken
des Mississippi und anderer in den
Golfstrom mündenden Flüsse, sind
durch die entsprechenden Farben
(blau und roth) abgegrenzt.

Presented by A. Petermann

ARKTISCHEN & A...

geographischer...

MEERE...

GROSSER ODER STILLER OCEAN

Kuro siwo Japanischer...

ATLANTISCHER OCEAN

NORD-AMERIKA

ASIEN

EUROPA

AFRIKA

ARKTISCHER OCEAN

1. COOK 1773.

3. BALLENY 1839

GÖTH...

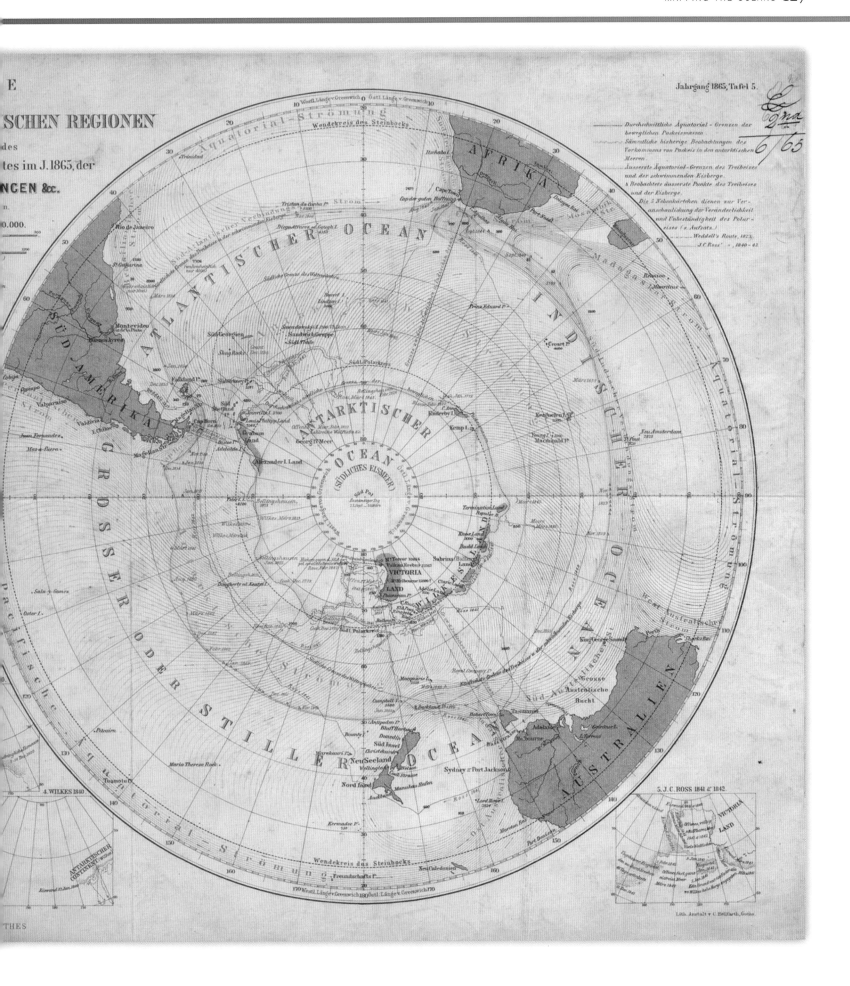

Above: The process of a bathymetric survey using the deep-sea sounding device developed by John Brooke. His inventions were used throughout the second half of the nineteenth century.

Right: In his later career, Fridtjof Nansen served as Norway's first Minister to the Court of St James and was professor of zoology and then oceanography at the University of Christiania.

Previous pages: A map of the Arctic and Antarctic regions produced by August Petermann, the "Sage of Gotha", in 1865 for the journal that would later be known as *Petermanns Geographische Mitteilunen*.

The next major step in mapping the ocean bottom came during the *Challenger* Expedition (1872–76), a major scientific effort funded by the British government and intended to answer comprehensive questions about the ocean. The scientific objectives included investigating the distribution of sea life at different depths; understanding the mechanisms by which the oceans circulate; determining the temperature, salinity, density and chemical composition of seawater at various depths; and investigating the topographical conditions of the seabed in the great ocean basins.

During *Challenger*'s programme, the scientists sailed more than 127,700 km (68,900 nautical miles), making stops at 362 sampling stations, where they collected many kinds of data, including a total of 492 soundings. These helped to outline abyssal plains, undersea mountain ranges and deep trenches in the world's oceans.

*Challenger* had been provided with the most up-to-date technology, some of which was relevant for mapping. This included a liquid-filled nautical compass developed by the American instrument-maker Edward Ritchie. With the damping provided by the liquid, together with multiple gimbals compensating for the ship's pitch, roll and yaw, the floating compass card remained relatively stable even in high seas. In the previous decade, Ritchie had also invented an improved marine theodolite, which had been used for the first precision surveys of American harbours and ports, and subsequently became the worldwide standard.

Sadly, *Challenger* did not carry a version of the deep-sea sounding machine that was developed by Sir William Thomson (later Lord Kelvin) after the start of the expedition. Rope had long been used for sounding in the ocean – *Challenger* was provided with 232 km (144 miles) of it for this purpose – because it had seemed impossible to manufacture strong wire in the lengths required. Thomson used steel piano wire, which could be wound mechanically, in conjunction with a small-diameter glass tube that was treated in such a way as to record pressure at depth. When the sounding element was removed from the water, the pressure recorded would indicate the depth of the ocean.

Although many of the world's coastlines and some of the ocean seabed had been charted by late in the nineteenth century, little was yet known about the Arctic Ocean, including whether there was land at the North Pole. For example, August Petermann, one of the world's most respected geographers of the century, had proposed that there was a polar land barrier extending from Greenland past the North Pole to Wrangel Island in the Russian Arctic. Others had suggested that the Eurasian half of the Arctic Basin contained an island-studded sea, much like the Canadian archipelago.

In 1893, Fridtjof Nansen, a Norwegian explorer and scientist already famous for being the first man to cross Greenland, set off to answer such questions. To demonstrate a transpolar current existed, he decided that he would deliberately enter the pack ice near the New Siberian Islands in his ship *Fram*, which was specially designed to withstand the pressure of the ice. He would then wait while the current carried him across the ocean near the North Pole to open water between Greenland and Svalbard. In general, Nansen's expedition went according to plan – until he left *Fram* to reach the Pole. He did not succeed and, having been separated from his ship, needed fortune on his side to be able to return safely. But Nansen had succeeded in changing the maps of the Arctic by proving there were no lands near the North Pole. Moreover, the soundings taken throughout the drift gave a more complete picture than ever before of the depths of the Arctic Basin.

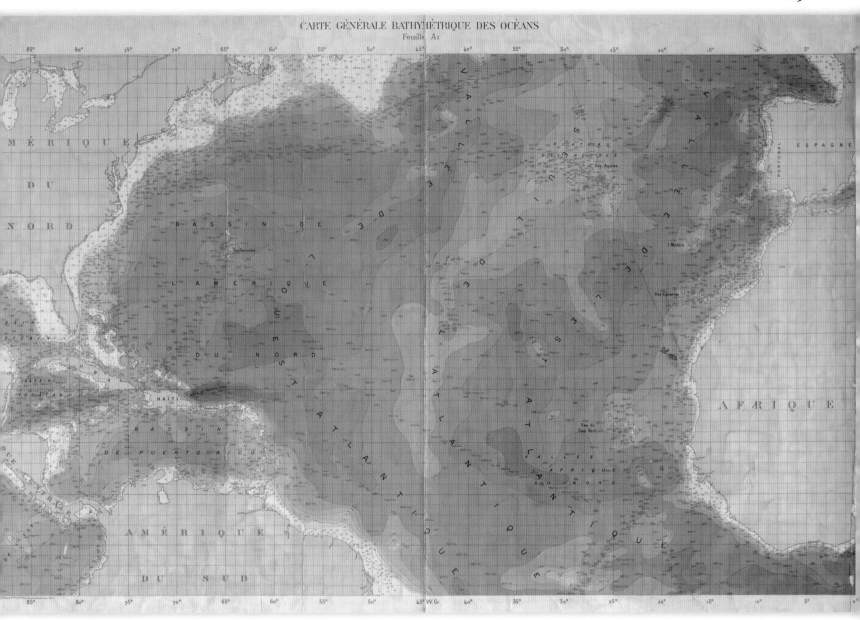

CARTE GÉNÉRALE BATHYMÉTRIQUE DES OCÉANS
Feuille A1

Three years after Nansen's return, a proposal at the Seventh International Geographical Congress led to the creation of a commission to prepare a bathymetric atlas of the oceans. In 1903 Prince Albert I of Monaco, one of the world's foremost oceanographers and marine cartographers, agreed to supervise the production of the *General Bathymetric Chart of the Oceans*. Published in 1905, the chart included 16 sheets of Mercator projections and eight of gnomonic projections (in which all great circles are shown as straight lines) at a scale of 1:10 million. The atlas was based on what seemed a huge number of soundings – 18,400 – but that number increased dramatically with the development of echo-sounding equipment following World War I.

There were some criticisms of the *General Bathymetric Chart of the Oceans* and a second edition was soon in preparation. In the meantime, another of the world's most respected oceanographers, Sir John Murray, produced a further major bathymetric map of the world. Murray had been a naturalist on *Challenger*, and when its scientific leader, Charles Wyville Thomson, died in 1882 while preparing the scientific reports, he had taken over as editor. He produced 50 volumes by 1896.

One of the first to note the existence of an extended Mid-Atlantic Ridge and of oceanic trenches, Murray's interests in bathymetry culminated in 1909 when he sponsored the most ambitious private oceanographic research cruise ever to have sailed at that point. Three years later, many of the cruise's findings were released in *The Depths of the Ocean*, published with the Norwegian marine zoologist Johan Hjort, who had been the expedition's scientific leader. Included in that work was Murray's Bathymetrical Chart of the Oceans Showing the Deeps, long since recognized as one of the classic works of bathymetry.

Above: A detailed bathymetric map of the North Atlantic Ocean, taken from the *General Bathymetric Chart of the Oceans*.

# The Polar Regions

Right: The South Circumpolar Chart produced following the circumnavigation of Antarctica by Fabian von Bellingshausen in 1819–21. Antarctica does not appear because not enough of it had been seen to either verify it was a continent or chart it as such.

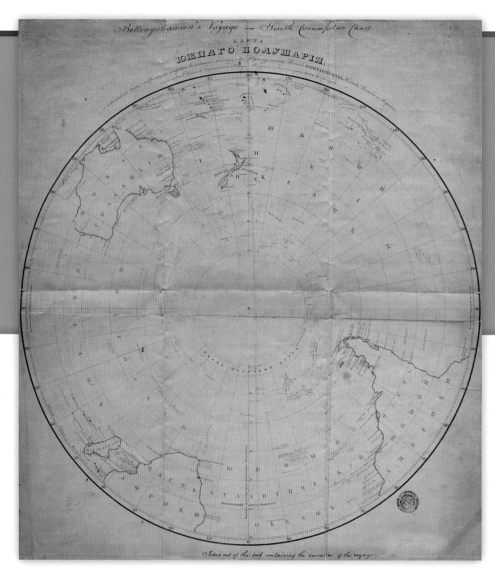

ALTHOUGH THE EUROPEAN ARCTIC, the Siberian coast, south Greenland and large segments of Rupert's Land had been mapped before the end of the eighteenth century, the high North American Arctic remained virtually unknown. However, the conclusion of the Napoleonic Wars left the Royal Navy with unemployed officers, men and ships, and in 1818, at the urging of John Barrow of the Admiralty, the first in a series of expeditions was launched to discover the Northwest Passage (see also page 84).

The Royal Navy officers who began to chart North America's Arctic coastline and archipelago would later be regarded as among the greatest polar explorers: John Ross, William Edward Parry, James Clark Ross and John Franklin. Within a decade, these and others had begun to penetrate the area's geographical mysteries, as shown by Franklin's map produced for the account of his second expedition.

Coincidentally, the year that Franklin left on his first expedition, the Antarctic also became a focus of exploration. In 1819 contrary winds forced the English merchant William Smith south in Drake Passage and he discovered the South Shetland Islands. Soon those islands and the nearby Antarctic Peninsula were being combed by sealers. However, because they did not want to invite competition, they did not announce or map their discoveries. Also in 1819, Fabian von Bellingshausen sailed south from Russia, and in the next two years he circumnavigated Antarctica. But his map could provide few certain details because heavy ice usually kept him far from the coast.

Two decades later, three major national expeditions entered the Antarctic and each produced maps from their voyages. Those of Charles Wilkes of the US Exploring Expedition (1838–42) contained far more coastline than the charts made during Jules-Sébastien-César Dumont

d'Urville's French expedition (1837–40), and were also the first to identify the new land as the "Antarctic continent". Unfortunately, James Clark Ross and John King Davis later sailed through areas that Wilkes had marked as land, obviously decreasing the usefulness of these charts.

Conversely, the "South Polar Chart" from Ross's expedition (1839–43) gave the most accurate accounting yet of the Antarctic, although his conclusion that "Wilkes's Land" did not exist caused considerable acrimony between the British and Americans. Ross's more detailed chart of Victoria Land showed the many discoveries he made after navigating through dense pack ice into the Ross Sea. The chart shows the mainland coast, Ross Island and a few other small islands, as well as the Great Ice Barrier (today known as the Ross Ice Shelf), which he traced for some 600 km (375 miles).

Ross retired from exploration after his Antarctic voyage, but in 1848 he agreed to lead a search for his friend John Franklin, who had disappeared in 1845 on an attempt to find the Northwest Passage. In the next decade, numerous attempts were made to find Franklin, but it was only in 1859 that Francis Leopold McClintock discovered that all

Above: Sir James Clark Ross was the first person to reach the North Magnetic Pole in the Arctic. Although he did not attain his goal of the South Magnetic Pole on his Antarctic expedition, he made numerous other discoveries.

Left: Otto Sverdrup accompanied Fridtjof Nansen on the first crossing of Greenland, captained the ship *Fram* on Nansen's famous polar drift and then took *Fram* to the Canadian Arctic as the leader of a hugely successful exploring and mapping expedition.

Right: Eric Marshall's map of the journey of the Southern Party on Ernest Shackleton's British Antarctic Expedition, 1907–09. Marshall's map was later found to be extremely accurate by the members of Captain Scott's Polar Party.

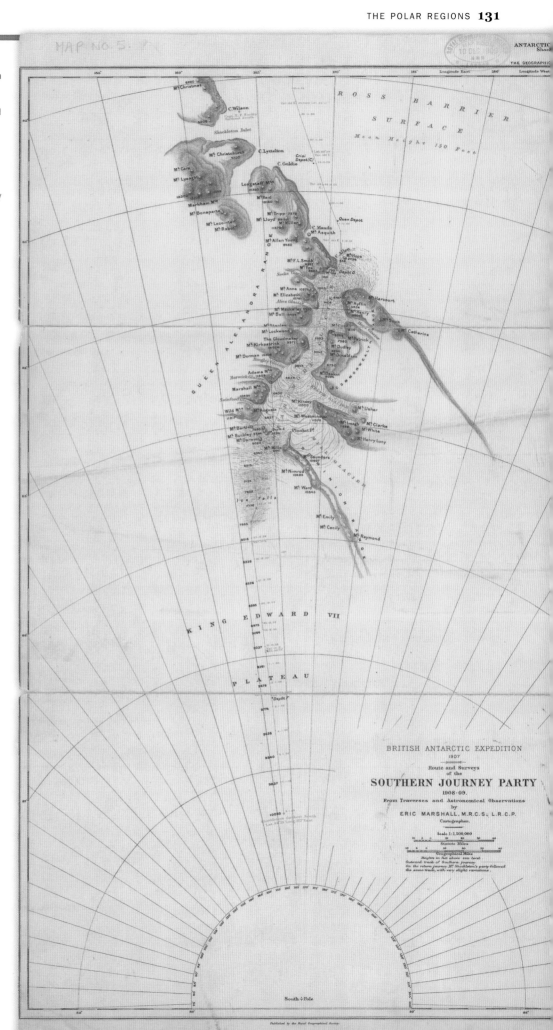

the members of the expedition had died in the high Arctic. A more positive outcome of the search was completing the charting of many of the islands and passages in the North American Arctic.

The last major step in completing the map of the Canadian archipelago was made four decades later. In 1898–1902, Otto Sverdrup led a Norwegian expedition to the region's northernmost islands. Sverdrup's party not only discovered the final three major islands – Axel Heiberg, Amund Ringnes and Ellef Ringnes – but it precisely mapped more than 260,000 sq km (100,000 sq miles), the largest area ever covered by a surface-based polar expedition.

Sverdrup's time in the Arctic coincided with the beginning of the "Heroic Age" of Antarctic exploration. A Belgian expedition under Adrien de Gerlache, Jean-Baptiste Charcot's French effort and Otto Nordenskjöld's Swedish scientific expedition each wintered in the vicinity of the Antarctic Peninsula and conducted extensive mapping. However, the complicated series of small islands and tiny straits to the west of the peninsula, along with the Larsen Ice Shelf and the mainland's glacial conditions, meant that none of the parties was able to produce a thorough map. It would not be until the British Graham Land Expedition (1934–37) that the peninsula was essentially correctly delineated.

Mapping was no easier on the far side of a continent that is more than 99 per cent covered by ice. There, the British National Antarctic Expedition under Robert Falcon Scott made the first serious attempt to discover if land lay in the interior. But it was not until 1908–09 that Ernest Shackleton crossed the Great Ice Barrier, ascended the Beardmore Glacier through the Transantarctic Mountains and reached the Polar Plateau. As shown by the route map prepared by surgeon and cartographer Eric Marshall, Shackleton's party reached a point only 157 km (97 geographical miles) from the South Pole. Beyond that, no map was needed, as the white, icy plains stretched unchanging to the Pole.

# Maps Reach a
# Wider Audience

**D**ESPITE THE EVER-INCREASING NUMBER AND TYPE OF MAPS, until well into the nineteenth century they remained primarily tools for mariners, explorers, military officers and government officials, or luxuries for the wealthy. But in the middle of the century, there was a rise in demand for maps. This was because of growing literacy rates; new processes in engraving, printing and distribution, which made maps affordable; and a growth in public interest in exploration and imperial expansion. At the same time, map-makers were establishing a stronger scientific base for cartography, achieving a higher quality and demonstrating more powerful academic, political, economic and social uses for maps.

Two organizations that led this growth of interest in maps were the Société de Géographie in Paris and the Royal Geographical Society (RGS) in London, founded in 1821 and 1830 respectively. These societies sponsored expeditions of discovery and published journals that detailed the findings and maps of the newly explored regions of the world. In the 1840s, many of the best maps for the RGS were drawn by the German cartographer August Petermann, who was appointed "Physical Geographer Royal" by Queen Victoria.

Petermann later moved to Gotha in Germany, and in 1855 he began producing a monthly geographical journal generally known as *Petermanns Geographische Mitteilungen*. He believed that it was only on a map that exploration could be clearly expressed. "That which, in a written report may be circumscribed or considered doubtful", he wrote, "must take a definite shape on the map." His *Mitteilungen* therefore emphasized maps

Above: August Petermann, about whom John Bartholomew later wrote: "no one has done more than he to advance modern cartography." Sadly, suffering from depression, Petermann committed suicide in 1878.

Below: A map of the route between Tabriz and Kaswin in Persia (modern-day Iran). Near the border of Azerbaijan, Tabriz was the scene of several military struggles between the Russians and Persians.

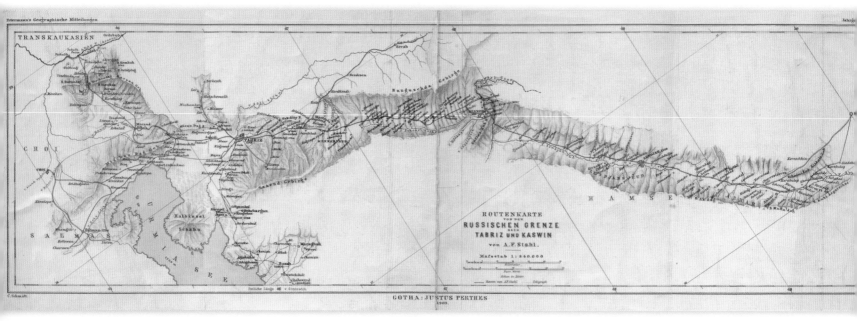

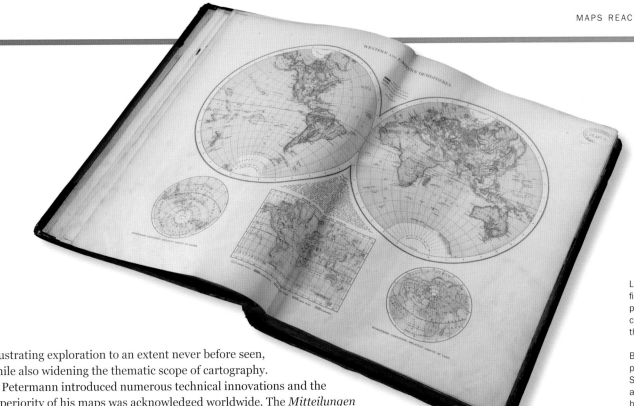

illustrating exploration to an extent never before seen, while also widening the thematic scope of cartography.

Petermann introduced numerous technical innovations and the superiority of his maps was acknowledged worldwide. The *Mitteilungen* became the leading organ of geography and exploration. In 1865 he assumed publication of *Stielers Handatlas*, an atlas produced in a "handy size" and sold at a moderate price. The "Sage of Gotha" – as Petermann was known – turned the *Handatlas* into the world's leading scientific atlas, which it was to remain for the rest of the century.

One of Petermann's assistants, John Bartholomew, also made a significant contribution to cartography. In Edinburgh, he popularized contour layer colouring, wherein altitudes are represented on a graduated colour scale, from white and grey at high elevation to dark greens at low. First introduced at the Paris Exhibition of 1878, this technique became standard cartographic practice.

Despite Petermann's pre-eminence, there was one issue upon which his viewpoint was not universally accepted: an initial or prime meridian. Cartographers had long used different initial meridians from which to measure longitudes on maps. Therefore, not only did British, French, Dutch and Italian maps show divergent coordinates for the same place, even more confusing was the fact that some countries used different prime meridians for sea charts from those used for land maps.

One of the earliest-established meridians was that of Antwerp, which Mercator used for his projection in 1569. Frankfurt, Rome, Madrid, Washington DC and St Petersburg were also used, as was Ferro (now El Hierro), the farthest west of the Canary Islands, which had been established by the French in the early seventeenth century because it was considered the most westerly part of the Old World. After Jean Picard laid down the Paris meridian, French cartographers adopted it as their prime meridian. Old maps often had a common grid, with Paris degrees at the top and Ferro degrees – offset by 20 from Paris – at the bottom.

The most widely used meridian (including by Petermann) was that of the Royal Observatory at Greenwich, particularly after its adoption by the Ordnance Survey. Numerous countries that used different meridians for land maps used Greenwich for sea charts. Although many geographers believed that a single prime meridian should be used, it was not until 1884 that an International Meridian Conference was convened at the urging of US President Chester A. Arthur.

Delegates representing 25 nations attended the conference, where Canada's Sir Sanford Fleming showed that 72 per cent of the world's shipping used Greenwich as the prime meridian. The final vote of the delegates was 22 for Greenwich, one against (San Domingo – now the Dominican Republic) and two abstentions (France and Brazil). The French had stated that they would only adopt Greenwich as the prime meridian if the British adopted the metric system; therefore, France retained the Paris meridian until 1914.

The argument to determine the prime meridian had little impact upon the public at large. But the process helped to unify the scientific basis of cartography and established geographic norms for map-making, as did the establishment of time zones.

The public's interest in geography continued throughout the century, and maps became increasingly fashionable. Numerous popular atlases appeared, including Blackie & Son's *Imperial Atlas of Modern Geography* in 1860 and 1872, and Paul Vidal de la Blache's *Histoire et Geographie: Atlas General* in 1894. Another was the 1895 publication of *The Times Atlas*, which included 173 maps and an index of 130,000 place names. *The Times* has continued to publish atlases ever since, with the most recent edition in 2010.

Left: A map of the world in the first edition of *The Times Atlas*, published in 1895. More than a century later, this remains one of the great atlases produced today.

Below: This structure marks the prime meridian in Greenwich. Straddling the centre line allows an individual to have a foot in both the eastern and western hemispheres.

# Thematic Maps

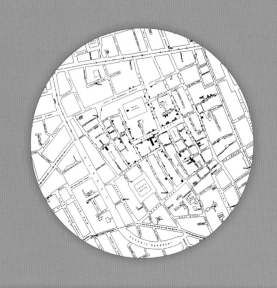

Left: William Smith's famous geological map was such a financial failure that he ended up in debtor's prison. After his release, he continued producing maps, including this New Geological Map of England and Wales in 1820.

Above: A section of John Snow's 1855 map plotting the previous year's cases of cholera in London. The clusters of the disease helped Snow identify the source of the outbreak as a public water pump.

**I**N A SENSE, ALL MAPS ARE THEMATIC because the information they present has been selected to show a particular range of characteristics, even if those are topographical features. However, some maps focus on physical, cultural, political, economic or other demographic subjects, and these are what cartographers consider to be "thematic" maps.

The roots of modern thematic mapping can be traced back to the seventeenth century, with the development of demographic statistics and theories of measurement and mathematical estimation. However, it was not until the nineteenth century that scientifically based data were presented in the form of maps about the world, its peoples and their activities, and the social environment.

Not surprisingly, one of the first thematic maps produced was about land use. Thomas Milne's map of 1800 not only covered a large area – 673 sq km (260 sq miles) – around London, but it was also considerably more elaborate, in terms of land classification, than any previously published.

Shortly thereafter, William Smith began work on his now-celebrated geological map, the Delineation of the Strata of England and Wales. Consisting of 15 sheets at a scale of just over three kilometres to the centimetre (five miles to the inch), it used an unusual system of colour intensity to distinguish sections of geological outcrops. It was published in 1815 by John Cary, London's foremost engraver; unfortunately, the giant map – 1.83 × 2.54 m (6 × 8.3 ft) – was a financial failure.

Four years before Smith's map appeared, George Cuvier and Alexandre Brongniart published their Carte géognostique, a geological map at almost the same scale as Smith's but covering a considerably smaller area. Although less scientifically sophisticated in its first edition than Smith's work, its pattern of identification by colouring became the standard.

Another new direction for mapping was public health. In New York in 1798, surgeon Valentine Seaman tried to demonstrate the source of yellow fever by charting the incidences in physical relation to the sewers in the city's docklands. Although Seaman's theory that odours from the sewers gave rise to the disease proved to be inaccurate, his work had produced the first disease map.

In 1833, Robert Baker's Sanitary Map of the Town of Leeds identified the squalid areas lived in by the "working poor". Then, in 1842, Edwin Chadwick, secretary of the English Poor Law Commissioners, produced his *Report on the Sanitary Condition of the Labouring Population*, which included maps that reinforced his arguments. Chadwick discussed at length the cholera outbreak of 1834, although he had no better idea than anyone else what had caused the outbreak. It was John Snow, a pioneer in medical hygiene, who eventually determined what had caused the spread of cholera. In 1855, Snow produced perhaps the most celebrated medical map of all time: he plotted the location of each house in London's Soho district in which a death from cholera had occurred, and from that he reasoned that the sources were specific water pumps. His deduction was confirmed by the decline in cholera when those pumps were closed.

By the middle of the century, thematic atlases were being produced. One of the most acclaimed examples was Heinrich Berghaus's *Physikalischer Atlas*, first published in 1838 with 90 maps on topics such as meteorology, climatology, geology, tides, magnetism, botany, zoology and ethnography. Berghaus worked closely with the Scottish geographer Alexander Keith Johnston, who in 1848 published *The Physical Atlas of Natural Phenomena*, which was essentially a condensed, albeit more lavishly illustrated, version of Berghaus's work.

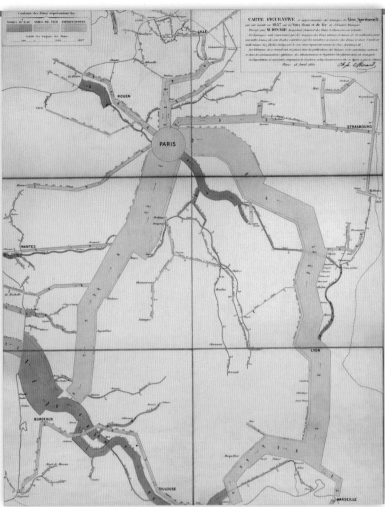

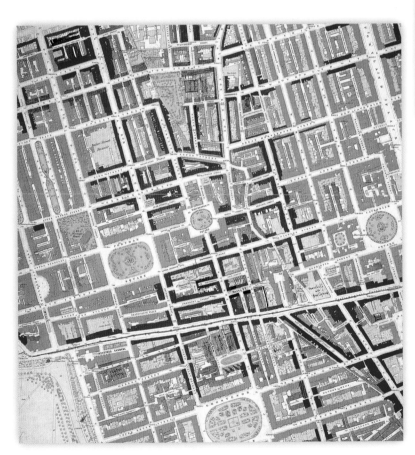

Demographic maps appeared in scholarly as well as popular publications. In 1849 Joseph Fletcher published several pioneering maps in the *Journal of the Statistical Society of London*. One was a "map of ignorance", for which he used marks – as opposed to signatures – made by men when registering for marriage, to show the proportion of illiterate men below and above the national average in the counties of England and Wales. He also produced maps of population distribution and of "persons of independent means".

One of the most creative cartographers of the second half of the nineteenth century was Charles Joseph Minard, a French civil engineer who, having retired in 1851 at the age of 70, then spent 19 years developing what he called *carte figurative* – flow maps and maps that used pie charts and other information display techniques. Through these he mapped dozens of topics, including imports of cotton and wool, the pattern of global emigration, the volume of tonnage shipped through European ports, the cattle sent to Paris for consumption, and what happened, and when, to Napoleon's men during the ill-fated invasion of Russia in 1812.

One of the greatest milestones in thematic mapping was the Descriptive Map of London Poverty, published in 1889 by social researcher and reformer Charles Booth, in conjunction with his study *Life and Labour of the People in London*. The result of a lengthy and intensive investigation, the map used different colours to classify seven financial ranges of the city's inhabitants. It was reprinted many times as Booth's book expanded to its classic 17-volume edition.

Above: A section of Charles Booth's Descriptive Map of London Poverty. This part of the city clearly contained a generally affluent segment of society, as yellow marks the wealthy, red the well-to-do middle classes and pink those with "good ordinary earnings".

Above right: Charles Joseph Minard's *carte figurative* detailing the tonnage of wine and spirits shipped throughout France in 1857. Pink indicates transport via railroads and green by boat. Paris was clearly the centre of the French transportation system.

# CHAPTER **5**

# Mapping in the *Modern World*

## (1914–2011)

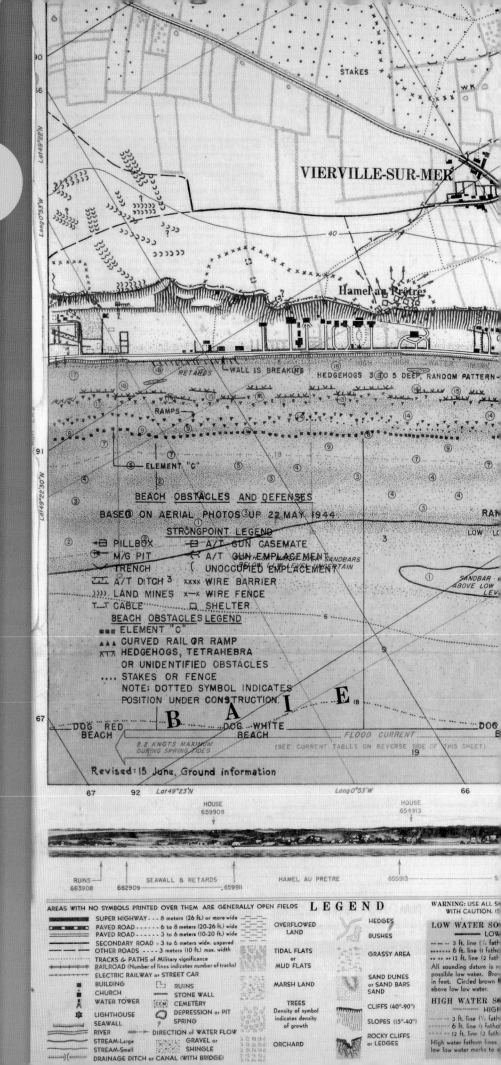

Right: Invasion map of the western part of Omaha Beach in Normandy. The plan for D-Day, 6 June 1944, was that the US 29th Infantry Division and additional companies of the US Army Rangers would assault the western half of Omaha Beach, which was further subdivided into sectors codenamed (from west to east) Able, Baker, Charlie, Dog Green, Dog White and Dog Red. The map is oriented with south to the top.

Gruchy

STAKES

le Haut Chemin

GRID SECTION

SPOT ELEV. 46 METERS

SPOT ELEV. 40 METERS

R.R. STATION

RETARDS

RAMPS

TOWER

SPOT ELEV. 33 METERS

STAKES

ENT "C"

ROCKS OCCUR THROUGHOUT

SANDBAR REPORTED ABOVE LOW LOW WATER LEVEL

EXACT LOCATION OF ROCKS AND POSITION OF L L W MARK IN THIS AREA UNCERTAIN

Pointe et Raz de la Percée

MAGNETIC NORTH 1944

VAR 9°55'W

WARNING

A TIDE FORMS HERE WHEN FRESH WINDS OPPOSE THE CURRENT. THE RACE EXTENDS 1/2 TO 1 MILE FROM POINTE ET RAZ DE LA PERCEE OVER SOUNDINGS OF 4 TO 8 FMS OR LESS. TURBULENT CURRENTS ARE TO BE ANTICIPATED IN THIS AREA UNDER ALL WIND CONDITIONS EXCEPT DURING SLACK WATER. THESE CONDITIONS MAY PROVE DANGEROUS TO LANDING CRAFT.

L A    S E I N E

CHARLIE    BEACH

EBB   CURRENT

1.2 KNOTS MAXIMUM DURING NEAP TIDES

(SEE CURRENT TABLES ON REVERSE SIDE OF THIS SHEET)

2.0 KNOTS MAXIMUM DURING SPRING TIDES

KNOTS MAXIMUM G NEAP TIDES

Long 0°54'W

93    65

64    Long 0°55'W    94

sur-Mer CHURCH
647910

VALLEY HOUSE
644913

HOUSE
645920

30' TOWER
638926

649917    647918
RETARDS

H CONTOURS

CONTOURS
RK
ine (3 fathoms)
ine (4 fathoms)
ine (6 fathoms)
level of lowest
water soundings
of beach in feet

CONTOURS
RK
ine (3 fathoms)
ine (4 fathoms)
ine (6 fathoms)
high high and

Panoramic shoreline sketch above shows beach as seen from water level height, approximately 2000 yards offshore, with sea at about 16 feet above low low water

## NOTE to COXSWAIN or NAVIGATOR

Building landmarks, especially near the beach, may be destroyed before any craft land. Terrain features, therefore, are much more reliable for visual navigation from panoramic shoreline sketch above. Green solid and broken lines with letter at each end on chart above refer to Beach Gradients on reverse side of this sheet. Also on reverse side are Sunlight and Moonlight Table, data on Inshore Currents and Tidal Stages.

# OMAHA BEACH-WEST (Vierville-sur-Mer)

Contours shown are at 10 meter (approx. 33 ft.) intervals above mean sea level, which is 13 ft. above low low water.

SCALE 1:7920 (8"=1 mile; 1"=220 yds.)

1 METER = 3.3 FEET

Map from GSGS 4490, sheets 79 & 80 and air photo examination. Grid square equals 1 kilometer (1000 meters).

METER SCALE    500    400    300    200    100    0    500    1000 METERS

YARD SCALE    500    400    300    200    100    0    500    1000 YARDS

GRID REFERENCE SCALE    0    1    2    3    4    5    6    7    8    9    10
This scale can be used to accurately measure tenths eastward or northwards in any grid square on the map.

LONGITUDE SCALE    0"    10"    20"    30"    40"    50"    60"

LATITUDE SCALE    0"    10"    20"    30"

These scales can be used to accurately measure seconds of latitude or longitude using latitude and longitude squares on the map.

6047

PREPARED BY COMMANDER TASK FORCE 122, APRIL 21, 1944

TOP SECRET - BIGOT    UNTIL DEPARTURE FOR COMBAT OPERATIONS—THEN THIS SHEET BECOMES RESTRICTED

# Politics and the Military

ORANGE FREE STATE

Showing Blockhouse lines
Blockhouses at Vaal Drifts (from October 1901)
& S. A. C. Posts
with dates of completion

The Blockhouses along the lines of railway were in most cases
more or less completed in July 1901; additional Blockhouses however
were built subsequently to strengthen the lines.

ONE OF THE ENDURING HISTORICAL THEMES of the twentieth century was the struggle for international political dominance. Throughout the period, mapping played an important role in the political propaganda of empires and other regimes, and it contributed a great deal to military efforts when open conflict occurred.

In France, a corps of *Ingenieurs-geographes* (geographical engineers) had been established in the eighteenth century to produce maps for the army and carry out topographical surveys. Napoleon reorganized the corps several times, and the engineers' primary duties became to prepare campaign maps, diagrams of fortifications and field works, and battle plans. Pre-battle mapping was so important to Napoleon that his chief cartographer, Louis Bacler d'Albe, was said to have been his most trusted advisor.

The first major conflict of the twentieth century – now called the South African War (1899–1902) – was fought between the British Empire and the Boer republics of the Transvaal and Orange Free State. Facing ongoing guerrilla warfare, the British suppressed the Boers by a series of measures that included moving civilians to concentration camps, conducting a "scorched earth" policy and establishing a series of blockhouses with miles of fence lines. Maps of this operation showed Britain's domination, but the actions led to widespread criticism of the British government, both at home and abroad, and were therefore a propaganda failure.

Not long afterwards, most of Europe was at war. As the fighting in France and Belgium evolved into a stalemate of trench warfare that epitomized World War I, maps showing the location of buildings, roads, rivers and forests became vital. The British and French armies initially used maps scaled at up to 1:120,000. By 1915 "trench maps" were produced at a scale of 1:10,000. By the late stages of the war, these were updated daily by French reconnaissance planes. (Aerial photography began in 1858 when Félix Nadar photographed parts of Paris from a balloon.) The maps of locations such as Ypres were so detailed that it has been argued that they contributed to overconfidence in the Allied military command.

In the years after World War I, maps were a powerful feature of political propaganda. Following the Bolshevik seizure of power in Russia in 1917, civil war broke out between the new government and a variety of opposition forces, including pro-monarchist elements, which were supported by Western countries. This foreign interventionism fuelled a long-lasting Soviet distrust of the West and encouraged propaganda such as that produced by the Russian artist Dmitri Moor, whose works included the 1921 map Be On Guard!, with its heroic Bolshevik soldier protecting the mother country.

In Germany, National Socialist (Nazi) propaganda also made full use of maps after Hitler came to power in 1933. Reich Minister of Propaganda Joseph Goebbels not only used the "big stage" – mass rallies and parades of party faithful – but reached out to ordinary Germans in popular forums. The Leipzig magazine *Illustrirte Zeitung*

Above: Map of the Orange Free State, showing blockhouse lines and South African Constabulary posts. By the end of the South African War, there were approximately 8,000 blockhouses of widely varying size and construction material.

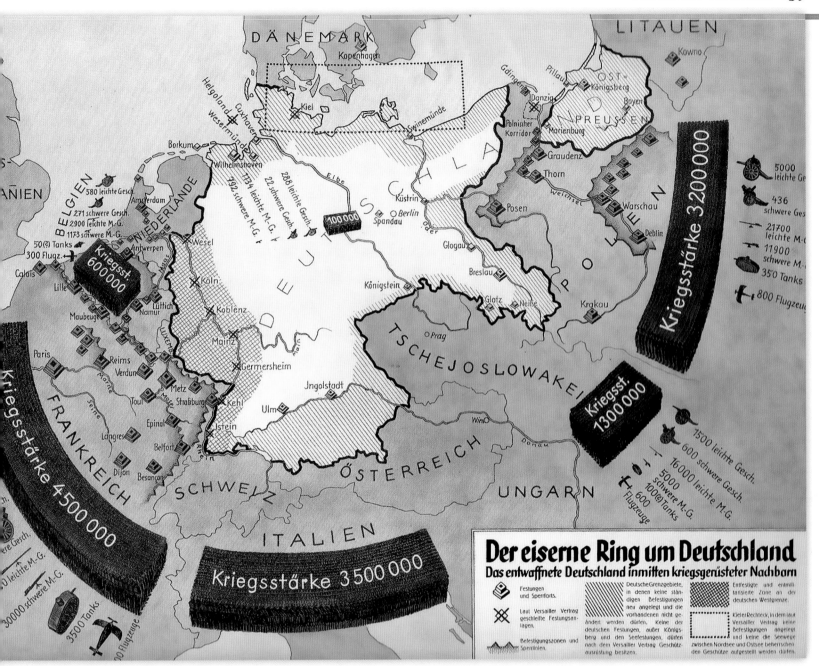

The map labels include:

DÄNEMARK, LITAUEN, Kopenhagen, Kowno, Königsberg, OST-PREUSSEN, Boyen, Kiel, Swinemünde, Danzig, Pillau, Gdingen, Polnischer Korridor, Marienburg, Graudenz, Thorn, Posen, Warschau, Küstrin, Berlin, Spandau, Oder, Weichsel, Deblin, Glogau, Breslau, Königstein, Glatz, Neiße, Krakau, Prag, TSCHEJOSLOWAKEI, Helgoland, Cuxhaven, Wesermünde, Borkum, Wilhelmshaven, Elbe, BELGIEN, Amsterdam, NIEDERLANDE, Wesel, Köln, Antwerpen, Koblenz, Calais, Lille, Lüttich, Namur, Maubeuge, Germersheim, Paris, Reims, Verdun, Mainz, Metz, Toul, Straßburg, Kehl, Epinal, Ulm, Jngolstadt, Langres, Belfort, Istein, Dijon, Besançon, FRANKREICH, SCHWEIZ, ÖSTERREICH, UNGARN, Wien, Donau, ITALIEN

**Kriegsstärke 3 200 000**

5000 leichte Ges.
436 schwere Ges.
21700 leichte M-G.
11900 schwere M-G.
350 Tanks
800 Flugzeuge

**Kriegsst. 1 300 000**

1500 leichte Gesch.
600 schwere Gesch.
16 000 leichte M.G.
5000 schwere M.G.
100(e)Tanks
600 Flugzeuge

**Kriegsst. 600 000**
580 leichte Gesch.
271 schwere Gesch.
2900 leichte M-G.
1173 schwere M-G.
50(e) Tanks
300 Flugz.

100 000

1134 schwere M-G.
22 schwere Gesch.
288 leichte Gesch.
792 schwere M-G.

**Kriegsstärke 4 500 000**

30 000 schwere M-G.
3500 Tanks

**Kriegsstärke 3 500 000**

**Der eiserne Ring um Deutschland**
**Das entwaffnete Deutschland inmitten kriegsgerüsteter Nachbarn**

Festungen und Sperrforts.

Laut Versailler Vertrag geschleifte Festungsanlagen.

Befestigungszonen und Sperrlinien.

Deutsche Grenzgebiete, in denen keine ständigen Befestigungen neu angelegt und die vorhandenen nicht geändert werden dürfen. Keine der deutschen Festungen, außer Königsberg und den Seefestungen, dürfen nach dem Versailler Vertrag Geschützausrüstung besitzen.

Entfestigte und entmilitarisierte Zone an der deutschen Westgrenze.

Kieler Rechteck, in dem laut Versailler Vertrag keine Befestigungen angelegt und keine die Seewege zwischen Nordsee und Ostsee beherrschenden Geschütze aufgestellt werden dürfen.

– a prestigious, high-quality and expensive weekly not dissimilar to *L'Illustration* in France or the *Illustrated London News* in Britain – used manipulated maps to stir up public fears about Germany's military vulnerability and the unfairness of the Versailles Treaty.

Maps for political purposes were not restricted to totalitarian regimes. In 1929–31 the explorer Douglas Mawson led the British, Australian, New Zealand Antarctic Research Expedition, the underlying purpose of which was to claim or strengthen previous claims to sectors of the Antarctic that the British and Australian governments did not want to fall under the control of Norway. Mawson's voyages and flights from his ship discovered and delineated large sections of coastline, while his maps laid the groundwork for the claims that became the Australian Antarctic Territory.

The outbreak of World War II led to developments in virtually every technical area of mapping. No project required greater secrecy in its gathering of vital data than the D-Day invasion on 6 June 1944. With British, Canadian and American forces making assaults on five sections of beach in Normandy – codenamed Sword, Juno, Gold, Omaha and Utah – it was necessary to map every imaginable aspect of the sea, coastline, inland, defences and obstacles that the invading forces might encounter.

In the years after the war, the world was still plagued by unrest, as countries struggled for independence, borders were redrawn and peoples were subjected to life under oppressive regimes. Maps often played a major role in such events. For example, in preparation for the partition of India in 1947, hundreds of maps were produced to provide different types of information for the

Above: A propaganda map showing the potential enemy forces surrounding Germany in the early 1930s. With size limits placed on the German armed forces by the Treaty of Versailles, the Nazis used this kind of map to frighten the German public.

Boundary Commission. Sadly, none of these maps provided a demarcation solution to the borders of India and Pakistan that could avert the violence following independence.

Similarly, although maps prepared for the Survey of Palestine under the British Mandate helped to establish the borders for the state of Israel, they did not guarantee a peaceful transition throughout Palestine when the UN Partition Plan of 1947 proposed a division of the region into different states.

Thereafter, throughout the Cold War period from the end of World War II up until the fall of the Berlin Wall in 1989, top-secret maps produced by both sides provided the bases upon which high-level decisions were made. For example, the CIA produced maps of Cuba that showed Soviet military placements and this information was used by President John F. Kennedy during the Cuban Missile Crisis of 1962.

Above: The 1921 Bolshevik propaganda map entitled "Be on Guard!" produced by Dmitri Moor. A soldier of the Red Army is shown defending his homeland from the Western invaders that supported the "White" forces in the Russian Civil War.

Right: A trench map of the area in which the Second Battle of Aisne took place in April–May 1917 during World War I. The terrible French losses in a frontal assault under heavy German fire at Chemin des Dames sparked a mutiny among the French troops.

151ST INFANTRY BRIGADE

149TH INFANTRY BRIGADE

8TH D.L.I.

6TH D.L.I.

4TH N.F.

6TH N.F.

BOUNDARY

DIVISIONAL

LEFT BRIGADE H.Q.

CENTRE BRIGADE H.Q.

RIGHT BRIGADE H.Q.

5TH D.L.I.

5TH N.F.

Pontavert

II.e Avril 1918

# Mapping
## for the
# Masses

THE PROCESS BY WHICH the general public began to obtain different types of maps began in the nineteenth century, and accelerated thereafter, so that today most households in North America or Western Europe own some kind of map.

More than anything else, it was an interest in travel that brought maps to the masses. Baedeker's travel guides had first been published in the 1830s, and their maps had earned praise from August Petermann. As new methods of transport developed, travel increased, and with it came new generations of maps.

In the years after the American Civil War (1861–65), the burgeoning US railroad system had so many carriers with different schedules and destinations that maps were critical for planning. Some of the largest railroad companies – including the Union Pacific and Central Pacific – issued national or sectional maps, and several notable publishers, such as Rand McNally, made fortunes from producing railway maps, timetables and guidebooks.

When safety improvements in the 1880s helped to make the bicycle popular, cycling maps quickly followed. Trying to provide something special, publisher George Philip produced waterproof maps, while Gall & Inglis published a Safety Cycling Map on which "unrideable roads" and dangerous hills were noted. In 1896 one of the finest cycling maps began to be produced, in seven parts: George Blum's *Cyclers' Guide and Road Book of California*. Each cycle road was shown in red and labelled with its condition and grade.

The advent of the car brought a need for road maps and travel information. In 1900 André Michelin published a guide about France, with maps to help travellers find food, accommodation and car assistance. Due primarily to its restaurant recommendations, the guide quickly became very popular.

Above: A Standard Oil road map of Chicago and its local environs, produced by H M Gousha Company in 1937. Free road maps are still available to members of the American Automobile Association.

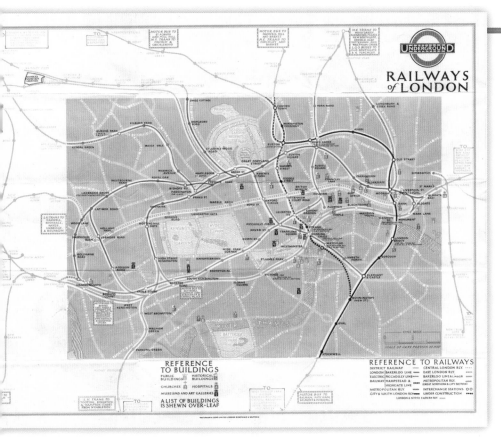

Nowhere was the need for road maps greater than the United States. In 1902 the American Automobile Association (AAA) was founded in Chicago, and three years later it published its first road maps, shortly after Rand McNally had expanded its mapping operation by venturing from the tracks to the roads. In the following decades, Rand McNally was one of three major producers of road maps and atlases. The company began to publish Auto Trails Maps – a series of nationally distributed regional maps – in 1917, and it even helped to establish the US's system of numbered roads. The other two major map publishers were General Drafting, which entered the business by producing road maps for the AAA, and H M Gousha Company, which was founded in 1926 by a former executive from Rand McNally. Gousha worked with the Continental Oil Company (Conaco) to develop the *Touraide*, a set of spiral-bound maps and directions with accommodation, restaurants and points of interest, ordered in advance and assembled individually for the traveller.

The oil companies did not take long to realize the profit to be made from Americans exploring the open road, so service stations soon began to distribute free maps. In 1920 Rand McNally started publishing road maps for Gulf

Above: A 1924 map of the Railways of London, in other words the London Underground. Designed by J C Betts, this edition was based on London above ground. For those familiar with the current schematic style, seeing the actual placement of the train lines is intriguing.

Right: Harry Beck's 1933 map of the London Underground. Although there are now more tube lines, and the Docklands Light Railway, the essence of the modern map remains the same as this classic version.

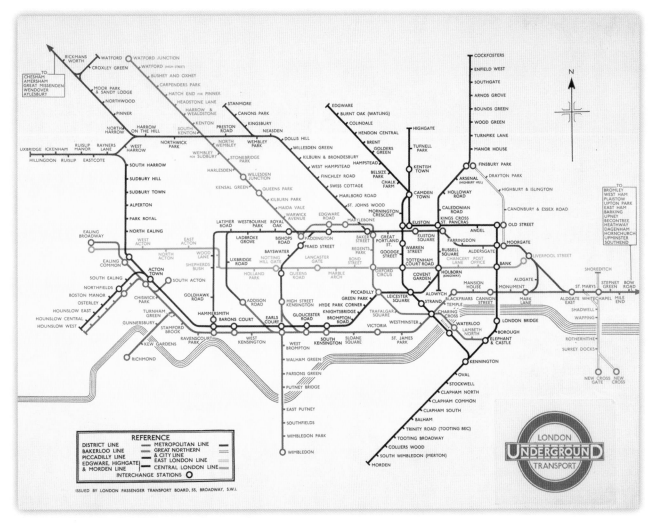

Overleaf: A railroad map of the United States, published by Rand McNally around 1880. As can be seen, there were a phenomenal number of lines criss-crossing the northeastern and north-central parts of the country.

Rand, McNally & Co.'s

NEW

OFFICIAL

RAILROAD MAP

OF THE

United States

AND

CANADA.

ENGRAVED AND PRINTED BY RAND, McNALLY & CO., CHICAGO.

Copyright, 1890, By Rand, McNally & Co.

MAP OF ATLANTIC COAST STATES ENGRAVED ON ENLARGED SCALE

Oil, and Gousha soon became the supplier for Standard
Oil of New Jersey (later Esso and Exxon). Free road maps
became part of the fabric of American life, and it has been
estimated that more than ten billion were distributed
before the 1970s, when the costs of the oil embargoes led to
the oil companies ending the practice.

Another travel-related map product was the aeronautical
chart for pilots. The first examples were produced in
France and England around 1911. Techniques progressed
greatly during World War I, and during the 1920s there
was continual improvement of maps for air navigation.
There were even efforts to establish international routes
by combining aerial and ground surveys. Sven Hedin's
Sino-Swedish Scientific Expedition (see page 71), which
began in 1926, was formed to establish an air route from
Berlin to Beijing. In 1930–31, the British Arctic Air Route
Expedition under Gino Watkins was established to survey
a shorter route between Britain and Canada.

New maps also became available for those who only
wanted to cross town. Some of the early maps of the
London Underground were based on the city above
ground; therefore, although they were accurate in terms of
distance and direction, the maps were confusing because
the stations in central London were so crowded together.
In 1931 Harry Beck, an electrical draughtsman, produced a
schematic similar to an electrical circuit, with straight lines
and the inclusion of only one feature above ground – the
Thames. The stations were also spaced relatively equally,
making the map much easier to read. Although Beck's map
was initially rejected as too radical, it was approved in
1933; he continued to refine it for the next 25 years.

Shortly after Beck's contribution to the mapping of
subterranean London, an equally significant achievement
was performed above ground. Phyllis Pearsall was a
painter who in 1935 became lost en route to a party in
London, due to the lack of a good map. This inspired
her to plot all of London, and the next year she traced
and catalogued its 23,000 streets. With cartographer
James Duncan, Pearsall then produced an atlas and a
comprehensive street index. Unable to interest any of the
major publishers, the two founded their own company –
the Geographer's Map Company Ltd – and produced what
was then called the *A to Z Atlas and Guide to London*. The
company still exists and now publishes more than 300
different A–Z maps and atlases.

Right: A three-dimensional axonometric map of the way that Los Angeles
would have looked in 1909. Pasadena and Glendale appear in the
distance to be sleepy, little towns, rather than part of the giant, bustling
megalopolis as they are today.

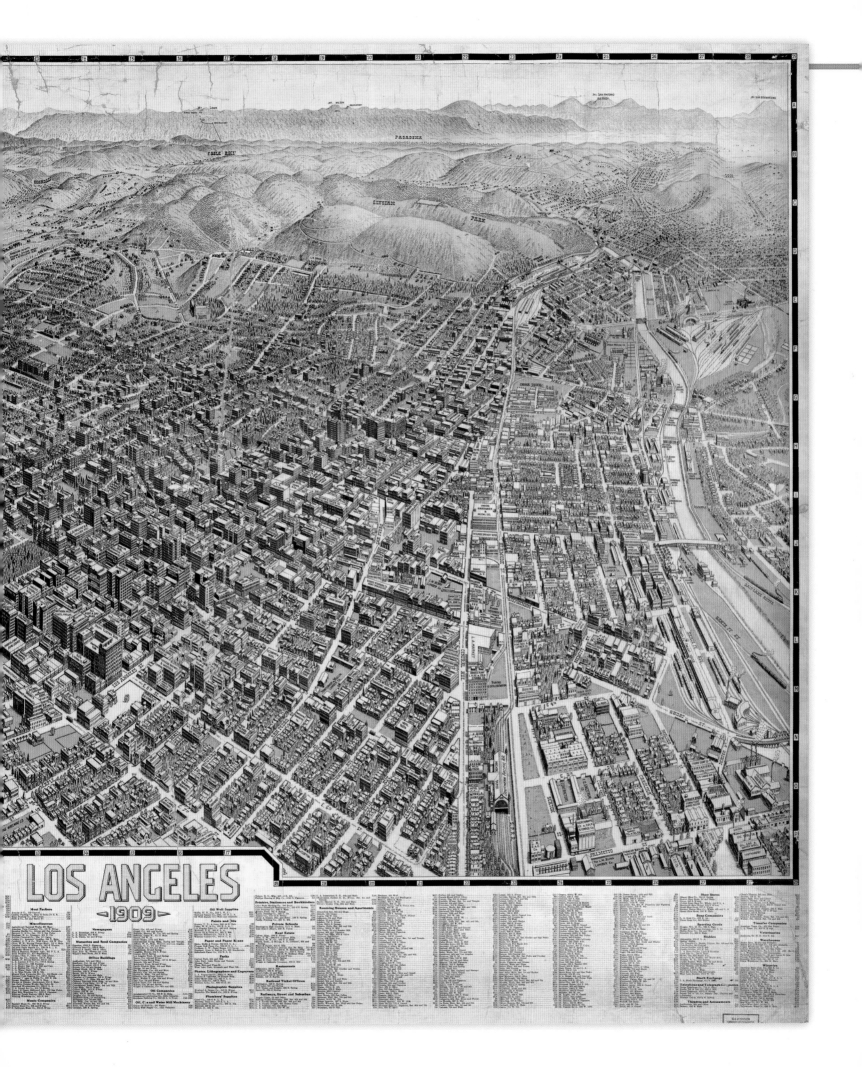

# LOS ANGELES
## ═ 1909 ═

# The Technological Revolution

LANL 15736

Left: Computer operators work on the Electronic Numerical Integrator and Computer by plugging and unplugging cables and adjusting switches. ENIAC contained 17,468 vacuum tubes, about 7,200 crystal diodes, some 70,000 resistors and approximately five million hand-soldered joints.

Below: A Landsat image of the northern Levant. Part of a geological survey of Israel, this map is a mosaic composed of 16 scenes obtained from Landsat 5, which was launched on 1 March 1984 and is still operational.

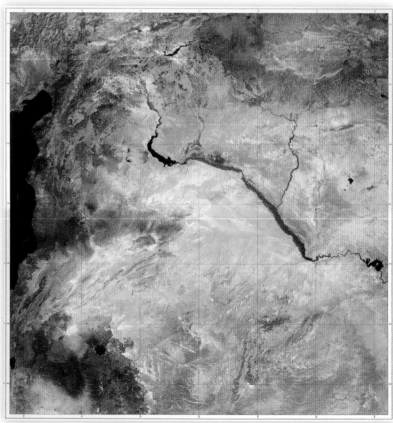

THE TECHNOLOGICAL REVOLUTION of the latter half of the twentieth century has had a profound impact on mapping. Many of the cartographic developments can be traced back to the US Army's need during the Second World War to produce tables that would project the correct trajectories required for large weapons to hit their targets.

In response, in 1946 engineers and scientists at the University of Pennsylvania created the Electronic Numerical Integrator and Computer (ENIAC), a U-shaped monster of a machine that was 24.4 m (80 ft) along its working front and included approximately 18,000 vacuum tubes. In the following years, ENIAC was used for calculations of atomic energy, studies of cosmic rays and to design wind tunnels. In 1950 it helped to generate the first computer-enhanced weather maps.

Since the 1970s, technology has advanced at an even more dizzying rate, totally transforming cartography. The widespread adoption of the personal computer, the creation of the internet and the ability of the private sector as well as governments to use satellite data have all changed the way that information is collected, managed and disseminated.

In 1972 the National Aeronautics and Space Administration (NASA) in the US launched the first civilian remote-sensing satellite as part of the Landsat programme (as it was to become known) to acquire imagery of Earth from space. In the following decades, six more Landsat satellites were launched (although Landsat 6 did not obtain orbit). The resultant images are not photographs but measurements of data in varying segments of the electromagnetic spectrum. These are obtained by remote sensing –

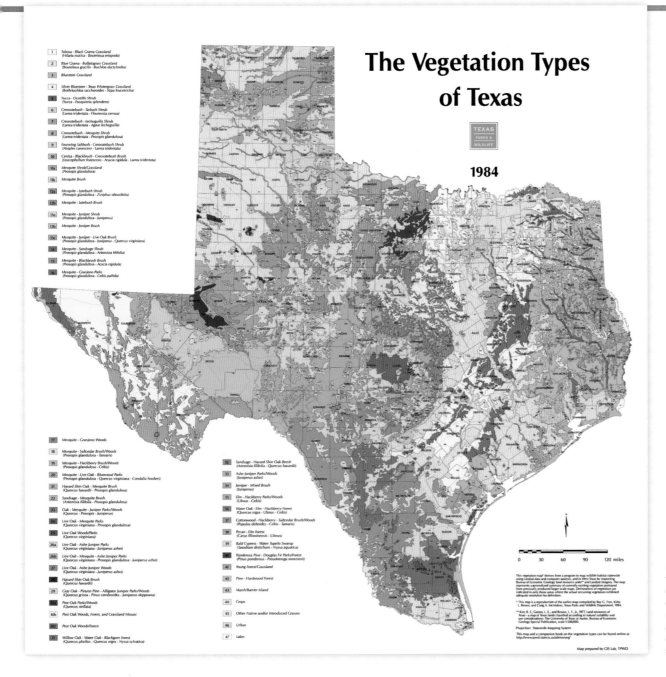

## The Vegetation Types of Texas

1984

Left: A GIS map of the vegetation types to be found in Texas, as of 1984. The vegetation was dominated by crops (light brown), mesquite mixed with other shrubs and brush, and live oak and post oak mixed with other trees and shrubs.

acquiring information about the Earth's surface by digitally recording reflected or emitted energy. Those data are then converted to images by computers, and the many colours in Landsat images signify a variety of information depending on the image.

Numerous other remote-sensing satellites have been put into orbit since Landsat. SEASAT (launched in 1978), TOPEX/Poseidon (launched in 1992) and Jason-1 (launched in 2001) have all compiled information about the Earth's oceans. Other satellites include ICESat (Ice, Cloud, and land Elevation Satellite), Radarsat (Canada's first commercial Earth observation satellite) and Terra, a satellite that orbits from Pole to Pole. Terra carries five remote sensors that every three months jointly collect almost 20 terabytes of data – an amount equivalent to that in the entire book collection of the Library of Congress.

In 1973 the global positioning system (GPS) was created by the US Department of Defense to improve upon earlier navigation systems. In the 1980s it became available for civilian use, and has since become the primary tool providing traditional mapping information – longitude, latitude and altitude – that previously took surveyors a considerable time to obtain precisely. Russia also has a global navigation satellite system – GLONASS – and the launch of the first four operational satellites for Galileo, the European system, is scheduled for 2011.

Equally as important as these methods of data collection is the way that geographical information systems (GIS) have changed the actual process of constructing maps. GIS integrates hardware, mapping software and digitized geographically referenced data to create, manage, analyse and display information in ways that reveal relationships, patterns and trends in map forms.

GIS has also transformed the final product of mapping, in that computer-generated versions tend to be virtual maps – temporary

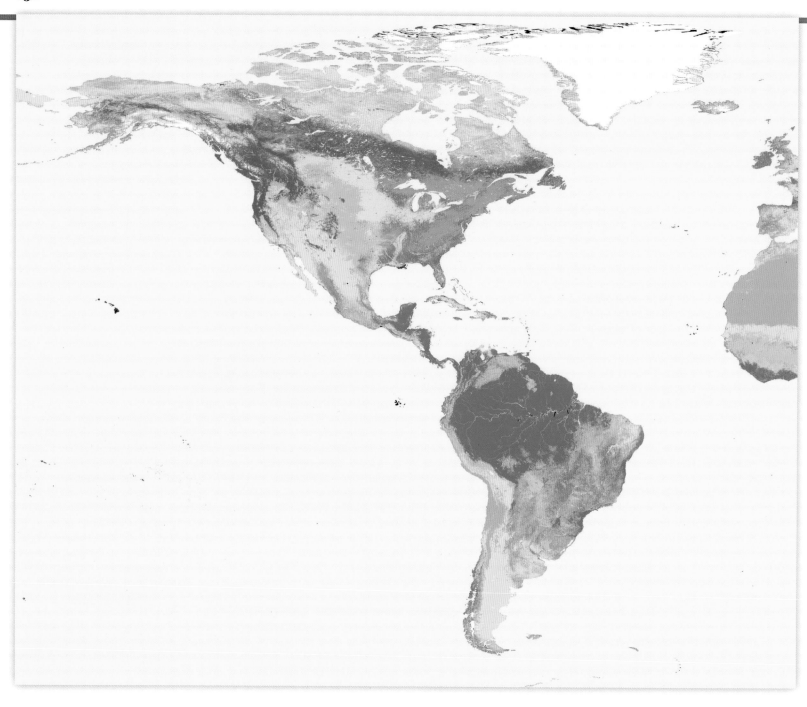

electronic images that exist only when on the computer screen (unless printed out). Although virtual maps can be vastly more pervasive than traditional ones – because there is a potential viewing audience of more than a billion internet users – they also have disadvantages. Like other information on the internet, there is little to guarantee the accuracy of such maps and their underlying data. Just as material available on the internet has not necessarily been peer-reviewed, maps may not have been appropriately vetted, making them of arguable value.

Recently, the three-dimensional axonometric map has become a valuable tool to commercial, industrial, architectural and governmental planners. Data are derived from a combination of ground-level photographs, orthophotographs (aerial photos geometrically corrected to have a uniform scale), satellite images and digital elevation models (DEMs), which are digital representations of the Earth's surfaces based on stereographic images, digitized topographic maps and laser scans. These are electronically manipulated to generate a bird's-eye view of a city and its buildings, trees and other aspects in a map that can be viewed from a variety of angles.

One of the latest developments in cartography is Google Maps, which was first announced in 2005. It is an application that produces street maps and an accompanying route planner for travel by car, public transport or foot from one location to another in a contiguous area. Depending on the region, it can also produce a satellite image, terrain features, photographs of the route and traffic information. It covers significant sections of the world with a Mercator-like projection, which does not function in the polar regions. A related mapping program – also introduced in 2005 – is Google Earth, which displays images of the Earth's surface in a variety of resolutions.

For further information, contact: RDHR Productions Ltd., P.O. Box 23312, Nicosia 1681, Cyprus. Tel: +357 2 676082; Fax: +357 2 677260; E-mail: info@rdhrproductions.com

Opposite: An image of North and South America produced from scans compiled from the Terra satellite. The Amazon rainforest, northern forests of Canada, agricultural lands in the centre of the United States and the less wooded areas of both continents are clearly defined.

Right: A Landsat image of the region surrounding Nazareth in northern Israel. A centre of Christian pilgrimage, the city of Nazareth is to the southeast of the area around the coastal city of Haifa and southwest of the Sea of Galilee.

Published March 2000 by Dr. John K. Hall, Geological Survey of Israel
GSI Report GSI/4/2000

NAZARETH

Cassini (Transverse Cylindrical) Projection
Scale 1:100,000. New Israel Grid Superimposed

# Mapping New Frontiers

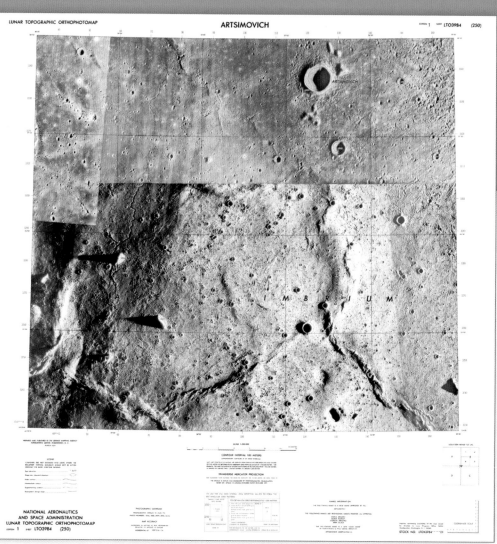

Right: A NASA lunar topographic orthophotomap of the region surrounding Artsimovich, a small crater located in the western Mare Imbrium on the Moon. Latin for "Sea of Showers", the Mare Imbrium is the second-largest basaltic plain on the Moon's surface.

As maps of the Earth's surface and continents have become more definitive, so too have cartographic efforts that relate to regions – or "worlds" – not seen so easily with the naked eye.

For more than a quarter of a century after World War II, Marie Tharp and Bruce Heezen of Columbia University's Lamont Geological Laboratory (now the Lamont-Doherty Earth Observatory) in New York worked together on the first systematic, comprehensive mapping of the ocean floor. Tharp drew bathymetric maps based on sonar readings made by Heezen and others on the Lamont research ship *Vema* and the Woods Hole Oceanographic Institution's *Atlantis*. In 1957 Tharp and Heezen released the Physiographic Map of the North Atlantic.

Twenty years later, Heezen and Tharp published the World Ocean Floor Panorama, which included details of the 65,000 km (40,000 miles) of mid-oceanic ridge system that they were the first to map. Their work proved essential for the eventual acceptance of the theories of plate tectonics and continental drift. In 1978, with the launching of SEASAT, data for oceanographic maps could come from space. SEASAT's mission lasted only 105 days, but it established a pattern of remote-sensing of the oceans that continues to this day.

Following Heezen's death in 1977, Tharp worked with Alvaro Espinosa and Wilbur Rinehart, who from 1960 to 1980 had plotted the location of more than 56,000 earthquakes. In 1981 the three produced a map of the seismic activity of the Earth. It clearly showed the relation of earthquakes, volcanoes and mid-ocean ridges, and documented the volcanic "ring of fire" that surrounds the Pacific Ocean.

In the 1950s, the mapping of the Moon took a major step forward when modern telescopes were linked with photography. In 1960 the *Photographic Lunar Atlas*, prepared by the US Air Force and the National Science Foundation and based on photographs taken from several major

observatories, was released as a portfolio of 230 photographs. Several supplements followed, including the *Orthographic Atlas of the Moon* (1960) and the *Consolidated Lunar Atlas* (1967).

However, lunar mapping was taken to unprecedented levels during the space race between the Soviet Union and the United States. In 1960 the Soviets produced the first map of the far side of the Moon, based on satellite photographs. The US responded with thousands of images taken during the *Lunar Orbiter* and *Apollo* missions, some of which were used to plan the *Apollo 11* Moon landing in 1969. The final *Apollo* missions included numerous high-quality aerial photographs, which were reproduced in the Lunar Topographic Orthophotomap (LTO) series published by the US Defense Mapping Agency in 1979. In this, the maps were produced at a scale of 1:250,000 using a transverse Mercator projection with contour levels at 100 m (328 ft). The series is still widely considered to contain the most accurate maps of the Moon's surface.

Right: Rebecca M Espinosa and Marie Tharp's 1982 map showing the seismicity (the frequency of earthquake activity) of the Earth based on data accumulated from 1960 to 1980.

Overleaf: One of the most significant maps of the past half-century was Marie Tharp and Bruce Heezen's World Ocean Floor Panorama of 1977, which, according to Tharp, "brought the theory of continental drift within the realm of rational speculation" and helped lead "to the more comprehensive theory of plate tectonics".

The logical step after charting the Moon was to extend mapping to the innermost planets. In 1972, *Mariner 9* orbited Mars and sent back 7,329 images, covering about 80 per cent of the planet and giving data about many surface features. This information was supplemented by high-resolution images obtained by the two *Viking* orbiting satellites and two landing units in 1976–82. Although such images tend to look like photographs, they are obtained via remote sensing and can be used by experts as up-to-date maps.

Between 1990 and 1994 the unmanned *Magellan* spacecraft mapped 98 per cent of the surface of Venus with synthetic aperture radar. This provided a photographic-quality, high-resolution series of images detailing craters, mountains, plains, ridges and other geographical features of the planet's surface. Many regions were imaged at different angles, enabling three-dimensional maps to be compiled.

Two decades after the *Viking* satellites reached Mars, the *Mars Global Surveyor* mission was launched to put a spacecraft into a low-altitude orbit, from where a wide-angle camera could document gullies and debris flows, producing new information that suggested liquid water had once

been present. Launched in December 1996, a month after the *Mars Global Surveyor*, the *Mars Pathfinder* mission included a landing unit that returned more than 16,500 images.

The Mars programme has continued into the twenty-first century. In 2001 the *Mars Odyssey* mission began using a Thermal Emission Imaging System, which enabled the mapping of surface and shallow-subsurface chemicals while concurrently collecting images in a variety of visual and infrared bands of the electromagnetic spectrum. These appear as photograph-like plots of the surface. The mission is currently still active.

The most recent mission to the red planet is the *Mars Reconnaissance Orbiter*, launched in 2005. Among several cameras aboard is the HiRISE (High Resolution Imaging Science Experiment), the most powerful camera ever to be used on a mission of planetary exploration. From its orbiting height of 260 km (160 miles), surface objects the size of a dinner plate can be identified. As the HiRISE focuses on small areas in unprecedented detail, the Context Camera provides wider-swath images of the terrain, which means that the data for a broader mapping perspective – with both small-scale and large-scale maps – are accumulated simultaneously.

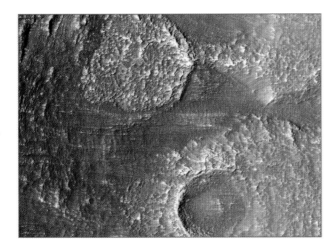

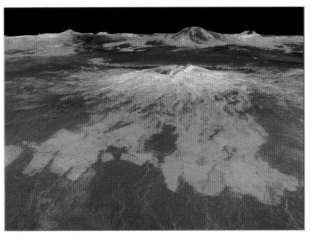

Far left: An image of the Lobate Debris Apron in the Deuteronilus Mensae region of the northern hemisphere of Mars. This HiRISE image was recorded in March 2010 and covers an area about 1 km (0.6 miles) wide, showing ice-rich materials thought to have been deposited about 10 million years ago.

Left: An image of the Sapas Mons volcano on Venus, as sent back to Earth from the Magellan – or Venus Radar Mapper – spacecraft. The giant volcano is approximately 400 km (250 miles) across. On the horizon in the back of the image is Maat Mons, the highest volcano on Venus.

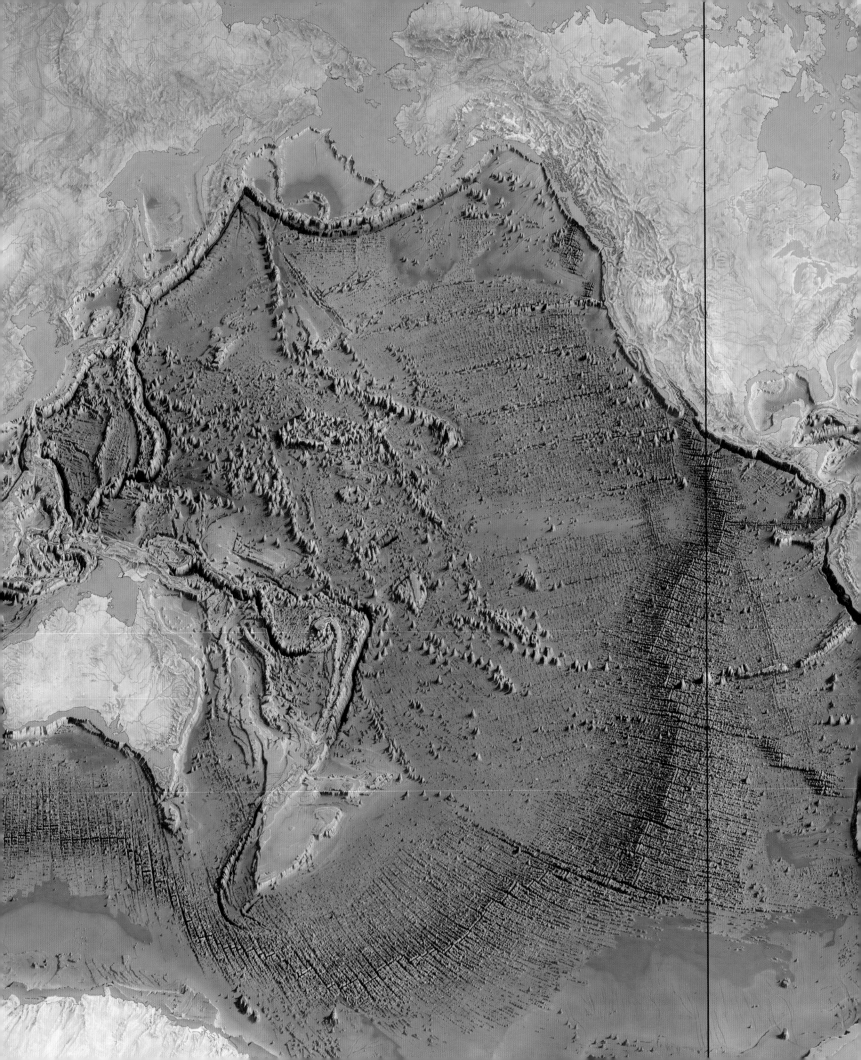

# The Controversies of Maps

Right: A statue of Piri Reis, located in Gelibolu on the Gallipoli Peninsula. The Turkish admiral, geographer and cartographer was responsible for one of the world's most controversial maps, which he produced in 1513. Now only a portion of the original survives.

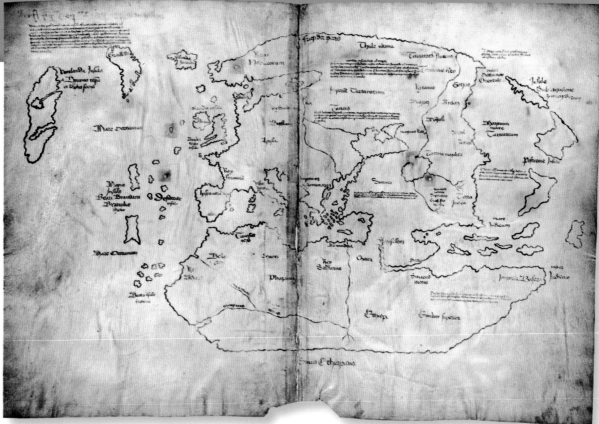

Left: The Vinland Map, which has been a source of continuing controversy for decades. It has been noted that the shape of Greenland is remarkably accurate, lending support to those who argue that it is not authentic, because there is no way that could have been known at the time. The debate looks to be nowhere near finished.

Opposite above: A reproduction of the Piri Reis map of 1513. The curving coastline in the southern reaches of South America has been interpreted as being the shore of Antarctica, although this has been disputed by many academic and scholarly authorities.

Opposite below: A world map produced for the United Nations using the Gall–Peters projection. The extremely elongated shapes of Africa and South America contrast greatly with those produced by the Mercator projection, which dominated Western mapping for centuries.

GIVEN THE HUGE NUMBER OF MAPS produced, it is not surprising that controversies have arisen regarding the provenance, age and accuracy of some. *Why* certain maps were created and what information the makers were trying to convey have also been contentious issues.

One of the most hotly contested cartographical controversies concerns the Vinland Map, a parchment chart of the Old World and the Atlantic Ocean, including, to the west of Greenland, an island identified as the "Vinland" of early Norse explorations. If the map – which is 41 × 28 cm (16 × 11 in) – is genuine, it demonstrates that mainstream western Europeans, and not just the Norse, were aware of North America before Columbus's voyages.

Donated to Yale University in the 1960s, the map was initially received with enthusiasm, but in 1974 analytical chemist Walter McCrone announced that studies of the map's ink revealed trace amounts of anatase (titanium dioxide) in a form that was only available after 1920. The map

was duly pronounced a forgery. However, McCrone's results were disputed by several scientists, including those who argued that anatase could have been produced as a byproduct in medieval times. Yale therefore pronounced the map to be authentic.

In 2002, two new studies announced results. First, the parchment proved to have a radiocarbon date of about 1434, the same age as a book with which it had been found. This raised questions about whether the map was equally ancient, or was a forgery made from blank pages taken from the volume. Concurrently, microprobe spectroscopy validated McCrone's earlier findings.

In 2004, the historian and cartographer Kirsten Seaver pronounced the map a modern fake, citing a number of reasons, including the ink, the wording on the map and the parchment source. But in 2009 further research resulted in a declaration that there was no reason to consider the

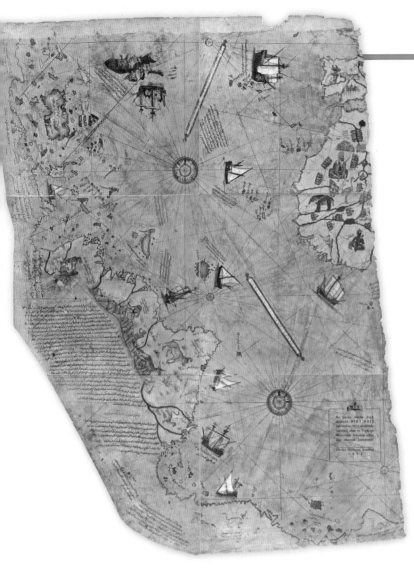

been many thousands of years before. This allowed Hapgood to reiterate his widely discredited theory that during a period long ago, before the Earth's poles had somehow shifted 15°, Antarctica (not then being at the Pole) had been ice-free. According to Hapgood, among Piri Reis's sources was a map produced by an ancient civilization that had charted Antarctica in its "pre-polar" period.

Hapgood's unorthodox notions were dismissed by most scholars, but they did find support from "alternative thinkers". In 1995 journalist Graham Hancock published *Fingerprints of the Gods*, which included Piri Reis's supposed Antarctic coastline as an argument for the notion of highly advanced ancient civilizations.

In reality, the Piri Reis Map bears little resemblance to Antarctica. And in answer to why the map-maker curved his coastline, many authorities agree that, as he approached the edge of his parchment, he simply turned the coastline to the east so that he could fit it in the available space.

Even something as seemingly innocuous as cartographic projections can be controversial. In 1973, Arno Peters, a German Marxist filmmaker, condemned the Mercator projection as "cartographic imperialism" and introduced what he called the Peters projection. This, he claimed, represented the size of all countries correctly and did away with Mercator's distorted emphasis on the Northern Hemisphere. Because, in an age of increasing concern about equality, this new projection gave more prominence to the developing world, many agencies and charitable organizations called for the Peters projection to be adopted universally.

However, much of the cartographic community remained unsupportive. Cartographers were disturbed because what Peters announced as his creation had been devised in around 1855 by the Scottish clergyman James Gall. They also condemned as false the assertion by Peters that his projection was the only "area-accurate" one. And they pointed out the extreme distortion not only in the polar regions but at the equator, making erroneous his claim of it being "totally distance-factual". Today, this controversy has mostly disappeared. Both the Gall–Peters and the Mercator projections have supporters and opponents, but, as many new projections have since become available, neither is used as widely as previously.

map a forgery – a conclusion that was again immediately disputed. Thus the debate continues, without consensus.

Unlike the Vinland Map, there is little question about the provenance of the Piri Reis Map. Drawn on gazelle skin in the style of a portolan chart, the map was produced in 1513 by a high-ranking officer in the Ottoman fleet, who is known as Piri Reis. The map's current size is 90 × 63 cm (35.4 × 24.8 in), but the surviving part, discovered in 1929 at the Topkapi Palace in Istanbul, is only a fragment of a larger, long-lost, world map.

The Piri Reis Map is thought to be the only one extant to derive directly from a chart produced by Christopher Columbus. Piri Reis's notes reveal that he had access to a wide range of information and compiled his map by drawing upon at least 20 existing Ptolemaic, Islamic and Western charts and *mappae mundi*, one of them being "the map of the western lands drawn by Columbus" – a map thought to have been made by Columbus on his second voyage.

The controversies relating to the Piri Reis Map started in 1966 with the publication of Charles Hapgood's book *Maps of the Ancient Sea Kings*. Hapgood claimed that where the coast of South America curved towards Africa at the map's bottom was not South America at all, but was instead a depiction of the ice-free coast of Antarctica as it had

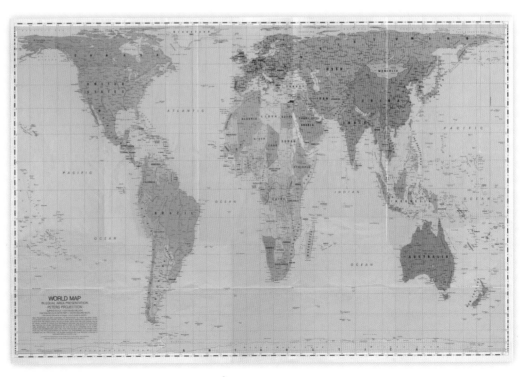

WORLD MAP
IN EQUAL AREA PRESENTATION
PETERS PROJECTION

# Index

Page numbers in *italic* type refer to illustrations or their captions.

# Further Reading and Research

## BOOKS

Akerman, James R. (Editor). 2006. *Cartographies of Travel and Navigation*. University of Chicago Press.

Akerman, James R. (Editor). 2009. *The Imperial Map: Cartography and the Mastery of Empire*. University of Chicago Press.

Bagrow, Leo, and R.A. Skelton. 2009. *History of Cartography*, Second Edition. Transaction Publishers.

Bendall, Sarah. 1992. *Maps, Land, and Society*. Cambridge University Press.

Berkeley, Edmund. 1974. *Dr John Mitchell: Man Who Made the Map of North America*. University of North Carolina Press.

Black, Jeremy. 1997. *Maps and History: Constructing Images of the Past*. Yale University Press.

Campbell, Tony. 1987. *The Earliest Printed Maps*. British Library.

Delano-Smith, Catherine, and Roger J.P. Kain. 1999. *English Maps: a History*. The British Library.

Dilke, O.A.W. 1985. *Greek and Roman Maps*. Thames & Hudson.

Edney, Matthew H. 1990. *Mapping an Empire: the Geographical Construction of British India 1765–1843*. University of Chicago Press.

Ehrenberg, Ralph E. 2005. *Mapping the World: an Illustrated History of Cartography*. National Geographic Society.

Harley, J. Brian, and David Woodward (Editors). 1987. *The History of Cartography: Cartography in Prehistoric, Ancient, and Medieval Europe and the Mediterranean*. University of Chicago Press.

Harley, J. Brian, and David Woodward (Editors). 1992. *The History of Cartography: Cartography in Traditional Islamic and South Asian Societies*. University of Chicago Press.

Harley, J. Brian, and David Woodward (Editors). 1994. *The History of Cartography: Cartography in the Traditional East and Southeast Asian Societies*. University of Chicago Press.

Harvey, Paul D.A. 1980. *The History of Topographical Maps: Symbols, Pictures and Surveys*. Thames & Hudson.

Harwood, Jeremy. 2006. *To the Ends of the Earth: 100 Maps That Changed the World*. David & Charles.

Howse, Derek, and Michael Sanderson. 1973. *The Sea Chart*. McGraw Hill.

Kish, George. 1984. *To the Heart of Asia: the Life of Sven Hedin*. University of Michigan Press.

Konvitz, Josef W. 1987. *Cartography in France 1660–1848*. University of Chicago Press.

Monmonier, Mark. 1991. *How to Lie With Maps*. University of Chicago Press.

Nebenzahl, Kenneth. 1990. *Maps from the Age of Discovery: Columbus to Mercator*. Times Books.

Ristow, Walter, W. 1985. *American Maps and Mapmakers: Commercial Cartography in the Nineteenth Century*. Wayne State University Press.

Robinson, Arthur H. 1982. *Early Thematic Mapping in the History of Cartography*. University of Chicago Press.

Shirley, Rodney W. 1983. *The Mapping of the World: Early Printed World Maps 1472–1700*. The Holland Press.

Skelton, R.A. 1972. *Maps: a Historical Survey of Their Study and Collecting*. University of Chicago Press.

Snyder, John P. 1993. *Flattening the Earth: Two Thousand Years of Map Projections*. University of Chicago Press.

Tooley, Ronald V., Charles Bricker, and Gerald R, Crone. 1976. *Landmarks of Mapmaking: an Illustrated Survey of Maps and Mapmakers*. Phaidon Press.

Wallis, Helen M., and Arthur H. Robinson (Editors). 1987. *Cartographical Innovations: an International handbook of Mapping Terms to 1900*. Map Collector Publications.

Walter, John A., and Ronald E. Grim (Editors). 1997. *Images of the World: the Atlas Through History*. McGraw Hill.

Wheat, Carl I. 1957–63. *Mapping the Transmississippi West*. 6 vols. Institute of Historical Cartography.

Wilford, John Nobel. 2000. *The Mapmakers*. Knopf.

Williams, Frances Leigh. 1963. *Matthew Fontaine Maury: Scientist of the Sea*. Rutgers University Press.

Winchester, Simon. 2001. *The Map That Changed the World*. Viking.

Woodward, David (Editor). 1975. *Five Centuries of Map Printing*. University of Chicago Press.

Woodward, David (Editor). 2007. *The History of Cartography: Cartography in the European Renaissance*. University of Chicago Press.

Woodward, David, and G. Malcolm Lewis (Editors). 1998. *The History of Cartography: Cartography in the Traditional African, American, Arctic, Australian, and Pacific Societies*. University of Chicago Press.

Yee, Cordell D.K. 1996. *Space & Place: Mapmaking East and West: Five Hundred Years of Western and Chinese Cartography*. Elizabeth Myers Mitchell Art Gallery, St John's College.

## PERIODICALS

*Cartographica*

*The Cartographic Journal*

*Imago Mundi: The International Journal for the History of Cartography*

*Journal of Cartography and Geographic Information Science*

*Petermanns Geographische Mittheilungen* [out of print]

## WEBSITES

British Library Map Room
www.bl.uk/reshelp/bldept/maps/maplibover/mapliboverview.html

David Rumsey Map Collection
www.davidrumsey.com/

Linda Hall Library
lhldigital.lindahall.org/

Map History / History of Cartography Virtual Library
www.maphistory.info/

National Library of Australia Maps
www.nla.gov.au/map/index.html

National Library of Scotland Map Collection
www.nls.uk/collections/maps

Newberry Library Map & Cartography Collections
www.newberry.org/collections/mapoverview.html

Princeton University Historic Map Collection
www.princeton.edu/~rbsc/department/maps/

Royal Geographical Society Maps & Atlases
www.rgs.org/OurWork/Collections/About+The+Collections/Maps+and+atlases.htm

Stanford University Historic Map Collections
library.stanford.edu/depts/spc/maps/index.html

UK National Archives Map Collections
www.nationalarchives.gov.uk/records/research-guides/maps-for-research.htm

US Library of Congress Geography and Map Reading Room
www.loc.gov/rr/geogmap/

University of Texas Perry-Castañeda Library Map Collection
www.lib.utexas.edu/maps/

Yale University Library Map Collection
www.library.yale.edu/MapColl/index.html

# Publishers' Acknowledgements

The Publishers would like to thank Alasdair Macleod, Head of Enterprise & Resources, and everyone at the Royal Geographical Society for facilitating this project. Special thanks are due to David McNeill, Map Librarian, and Julie Cole, Collections and Enterprise Assistant at the Royal Geographical Society for all their time and effort in providing advice and answering queries.

## PICTURE CREDITS

The publishers would like to thank the following sources for their kind permission to reproduce the pictures in this book.

Key, t: top, b: bottom, l: left, r: right.

**Akg-Images:** 38 tl, /Album / Oronoz: 34 tr, 48–9, /Anonym / Imagno: 131, /IAM/World History Archive: 70 t, /North Wind Picture Archives: 58 tl

**Alamy Images:** /John Farnham: 133

**British Library:** 62–63, 139, 140

**British Museum Photographs:** /© The Trustees of the British Museum: 17 tr

**Canadian War Museum:** 138

**Cartographic Images:** 10, 12 b, 16 t, 17

**Corbis:** 68 l, 123, 148, /Jon Arnold / JAI: 13, / Mimmo Jodice: 11 b, /Lee Snider/Photo Images: 78, /The Gallery Collection: 71, /Underwood & Underwood: 34 tl

**David Rumsey Map Collection:** /www.davidrumsey.com: 79, 105, 106, 107

**Department of Commerce:** 149

**Getty Images:** /Dorling Kindersley: 65b, /Lanz von Horsten: 38 b, /SSPL: 70 b, 72 t, 90 tl, 90 tr

**Imagestate:** 156 r

**Library of Congress:** 43, 82–3, 96, 102–03, 125 r, /© 1982 Rebecca M. Espinoza and Marie Tharp: 153, 154–5, /Geography and Map Division: 16 b

**London Transport Museum:** /TfL from the London Transport Museum Collection: 143 t, 143 b

**NASA:** 14, 150, 152, 153

**NOAA Photo Library:** /National Oceanic and Atmopheric Administration/Department of Commerce: 124 r, /National Oceanic and Atmospheric Administration/Department of Commerce: 128 l

**National Archives:** 136–37, 140–41

**National Library of Australia:** /(RM 750): 75

**National Maritime Museum:** 73

**Oregon Rock Art:** /www.oregonrockart.com / D. Russel Micnhimer: 8 tr

**Picture Desk:** 150, 153 t, 153 bl, 153 br, /Egyptian Museum Turin / Gianni Dagli Orti: 9, /Kharbine-Tapabor / Boistesselin: 18, /Museo Mexico / Collection Dagli Orti: 41, /Musée des Antiquités St Germain en Laye / Gianni Dagli Orti: 8 tl

**Private Collection:** 17 b, 134 r

**Royal Geographical Society:** /John Arrowsmith: 112–13, /Johan Blaeu: 54–55, /Roy Fox: 118, /George Philip & Son Ltd: 51 r, /James Rennell: 87, /W. Smith: 134 l, /J.T. Walker: 104 r, / David Wilson-Barker: 19 br

**Scala Archives:** /Florence / courtesy of the Ministero Beni e Att. Culturali: 39 t

**Stanford University:** 12 t

**The Bridgeman Art Library:** /Archives de l'Ecole Nationale des Ponts et Chaussées, Paris / Archives Charmet : 135 r, /Palazzo Ducale, Venice, Italy / Alinari : 35, /Private Collection / Photo © Bonhams, London, UK: 56t, /Scott Polar Research Institute, University of Cambridge, UK: 130, / Virginia Historical Society, Richmond, Virginia, USA: 124 l, /© Museum of London, UK: 135 l

**Topfoto.co.uk:** /Alinari: 47, /The British Library / HIP: 19r, /The British Library /HIP: 59, /The Granger Collection: 11 t, 68 r, 97, /The Print Collector/HIP: 101 t, /UK City Images: 58 b, / World History Archive: 8 b, 20–1

**Yale University Press:** 156

Every effort has been made to acknowledge correctly and contact the source and/or copyright holder of each picture and Carlton Books Limited apologises for any unintentional errors or omissions which will be corrected in future editions of this book.